Studies in History and Philosophy of Science

Volume 52

More information about this series at http://www.springer.com/series/5671

Ian Wills

Thomas Edison: Success and Innovation through Failure

 Springer

Ian Wills
School of History and Philosophy of Science
University of Sydney
Sydney, NSW, Australia

ISSN 0929-6425   ISSN 2215-1958   (electronic)
Studies in History and Philosophy of Science
ISBN 978-3-030-29939-2   ISBN 978-3-030-29940-8   (eBook)
https://doi.org/10.1007/978-3-030-29940-8

This Springer imprint is published by the registered company Springer Nature Switzerland AG.
The registered company address is: Gewerbestrasse 11, 6330 Cham, Switzerland

Acknowledgments

My profound thanks to Ofer Gal whose many suggestions and encouragement (and occasional harassment) made this book possible. Thanks too are due to Paul Israel, David Miller, and Paolo Palmieri who read and commented on the previous version of this book. Finally, I want to thank my wife, Dale, who has shared me with Thomas Edison for years.

Contents

Chapter 1
Introduction

> Before I am done with it, I mean to succeed. I have the right principle and am on the right track, but time, hard work and some good luck are necessary too — It has been just so in all my inventions. The first step is an intuition, and comes with a burst — Then difficulties arise. This thing gives out then that. "Bugs" as such little faults and difficulties are called, show themselves — Months of intense watching, study and labour are required before commercial success — or failure — is certainly reached.[1]

Thomas Edison wrote this in a letter to an acquaintance in 1878, soon after starting work on his electric lighting system. It is a telling description of his way of inventing because his laboratory notebooks are filled with descriptions of bugs – things that "gave out" – and the ways in which he overcame them. Edison used the word bug for these problems so early and so frequently that he probably coined it. As early as 1873 he described a "bug trap" that overcame a bug in one of his telegraph inventions, 16 years before the earliest citation in the Oxford English Dictionary (1889), also from Edison.[2,3]

Bugs were a constant problem for Edison and his co-workers. One of them, Edison's 16 year old nephew Charles, drew Fig. 1.1 apparently out of frustration with bugs.[4]

This is not just an etymological anecdote because bugs, those "little faults and difficulties", were Edison's constant companions. Each bug, even if a "little fault", meant the failure of an invention to work as intended. Each bug, if ignored, could have rendered an invention a failure.

Failures they may have been, but they were not disastrous. Dealing with them and, at times, exploiting them, were crucial to Edison's way of working. Over his

---

[1] TAEB 4:1570.

[2] TAEB 2:348.

[3] Third edition Oxford English Dictionary, "Bug, N.2," (Oxford University Press, 2006). http://www.oed.com/

[4] Charles ('Charlie') Pitt Edison (1860–1879) Thomas Edison's nephew and promising inventor. After the death of James Adams, Charles Edison followed him to Paris where he too died, aged 19.

© Springer Nature Switzerland AG 2019
I. Wills, *Thomas Edison: Success and Innovation through Failure*, Studies in History and Philosophy of Science 52, https://doi.org/10.1007/978-3-030-29940-8_1

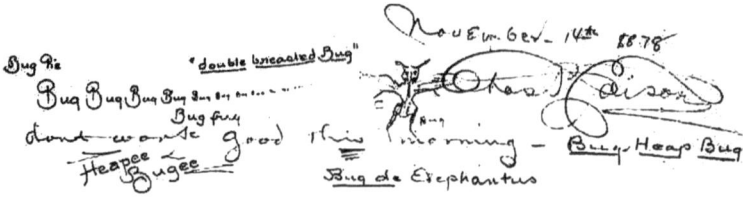

Fig. 1.1 Charles Edison's image of a bug (TAED NV18:84)

lifetime, Edison's attitude to bugs and failure generally was overwhelmingly positive. He could even be described revelling in them. A bug was not the ruinous end to a venture but a puzzle to be solved and perhaps an opportunity for new directions.

Despite many notable successes, Edison's career was also marked by many failures. The majority were minor failures, bugs, and known only to Edison and the people around him. After the failure of a crucial experiment, Edison commented that at least he had learnt that "the thing couldn't be done that way".[5] Edison certainly had learned that, but failures did more than just point to things that did not work. Indeed, he used failure so extensively and in so many innovative ways that his approach to it was fundamental to his prodigious inventive output. Failure was so significant to Edison's way of working that it is possible to follow his successes by following his failures. Chapter 2 relates how an embarrassing failure when demonstrating his automatic telegraph in 1873 led him to invent a variable resistor using carbon granules. The resistor in turn also failed to work as Edison intended but the cause of its failure, the sensitivity of carbon granules to vibration, became the basis for Edison's invention of the carbon microphone, a component crucial to the initial success of the telephone and used in most telephones for the next century.

Edison's fame comes from his public successes but a few of his failures were significant and very public, such as his claim to have discovered a new force of nature (Chap. 8) and the failure of the Naval Consulting Board, which he led during World War I, to produce any significant inventions for the war effort. Perhaps Edison's greatest failure was financial. This was his unsuccessful attempt to enrich iron ore using electromagnets. Edison lost the equivalent of 300 million dollars in today's monetary value in the venture but, turning failure into success, he adapted some of his ore processing inventions it the mass production of Portland cement, the basis of concrete. In so doing, he not only transformed financial failure into a commercial opportunity but also revolutionised construction in the twentieth century.

Thomas Alva Edison may have been born in obscurity in Ohio in 1847 but by the age of 31 was he was being described on both sides of the Atlantic as a genius and a

---

[5]Thomas A Edison, *The Diary and Sundry Observations of Thomas Alva Edison*, ed. Dagobert D Runes (New York: Philosophical Library Inc., 1948), 43.

magician.[6,7] By the time of his death in 1931 he had become a national treasure symbolising American ingenuity and inventiveness. Edison applied for his first patent in 1868 at the age of 21 and his last in 1931 aged 84. He remains America's most prolific inventor with almost 1100 patents to his name.

Edison received only a few months of formal education (he was largely home schooled by his mother) and began work at 12 selling newspapers on the local railroad. In his teens, he worked as an itinerant telegrapher, inventing in his spare time but by 22 was a full time inventor. His inventive output was prodigious, peaking at over 100 patents in 1 year. Initially ridiculed for making extravagant claims, Edison came to be known for producing near-miraculous inventions like the Phonograph, electric lighting and motion pictures. He was the Wizard of Menlo Park.

Early in his career Edison learnt that to protect his inventions from competitors, he needed detailed records of his progress in developing them. These records served as evidence in possible (often inevitable) future patent litigation. This practical need resulted in the accumulation of a huge amount of documentation. Edison's celebrity meant that that the documents were preserved after his death and are currently being published progressively by the *Edison Papers* project at Rutgers University.[8]

My interest in Thomas Edison began with a chance encounter with the book edition of the *Edison Papers*.[9] Leafing through the pages I was surprised to find that I could recognise the thought processes recorded in his laboratory notebooks. They were processes I was familiar with as an engineer developing new designs. Here were the kind of successes, failures, musings, discussion of options, blind alleys followed, mistaken assumptions, and so on, that are part of the process of creating something new.

Apart from being the papers of an exceptional person, the papers themselves are exceptional because they preserve these processes. The norm, in my experience, is that while similar records of conventional projects may be held in for a few years or even archived, almost all are discarded over time. Moreover, while practitioners in most areas including engineering and scientific research may be successful at what they do, they tend to be poor at describing how they did it. The result is that their retrospective explanations are often inaccurate and incomplete, particularly in relation to problems (bugs) overcome and even the recollection of why past difficulties were even difficulties.

In contrast, the *Edison Papers* preserve an easily accessible gold mine of material that can be used to study Edison's approach to inventing and by extension, the

---

[6]Washington Post. "Genius before Science." *Washington Post*, 19 April 1878.

[7]TAED MBSB1:171.

[8]Thomas A. Edison Papers. 2019. "Thomas A. Edison Papers." [web site]. The Thomas Edison Papers, Rutgers, The State University of New Jersey. http://edison.rutgers.edu/

[9]Thomas A Edison, *Menlo Park: The Early Years, April 1876–December 1877*, ed. Paul B Israel, Keith A Nier, and Louis Carlat, vol. 3, The Papers of Thomas a Edison (Baltimore: Johns Hopkins University Press, 1989).

process of innovation more generally. The papers show Edison inventing as he did it, rather than what he remembered decades later. This book contains processed ore from the *Edison Papers* gold mine. While previous accounts of Edison's approach to inventing, including his own, acknowledge his relentlessly positive attitude to failure, they have not addressed the significance of failure both to his success as an inventor and its role in the breadth and number of his inventions. Examining Edison's failures also highlights the importance of failure to innovation more generally; for those who seek to produce something that is both new and successful must be serious in their approach to failure.

This book is divided into four parts examining Edison at work and extending the conclusions drawn from Edison at work to issues of innovation more generally.

1.1 Part I: Edison and Failure

Part I focuses on Edison's use of failure to create inventions. It begins with a detailed examination of Edison's path to one of his most important but lesser known early inventions, the carbon microphone. This invention was important because it transformed Alexander Graham Bell's rudimentary telephone into a practical system, capable of transmitting sound over hundreds of kilometres rather than just between rooms.[10] Edison's development of the carbon microphone illustrates the many ways he used failure in addition to eliminating alternatives. Crucial to this process is the identification of success criteria, that is, the criteria used to judge whether or not something succeeds or fails.

Examples taken from Edison's development of the carbon microphone are used to develop a theoretical approach to failure and success that is applicable to not only to Edison's successful inventions, but to success and failure generally.

Part I then turns to the question of systems, developing the concept of the functional system as a way of understanding inventions and innovation generally. (A functional system is a collection of components that interact and are related by a

---

[10]Alexander Graham Bell (1847–1922) Bell was a Scottish-born inventor-entrepreneur who initially followed his father in becoming a teacher of the deaf. Bell migrated to Canada, later moving to the United States, there taking as a student Mabel Hubbard, the deaf daughter of Gardiner Hubbard a prominent Boston lawyer and financier. Bell had been experimenting with automata that produced speech and with telegraphy for some years and at Gardiner Hubbard's suggestion developed his own ideas for the telephone. In 1877 Bell married Mabel Hubbard and in the same year, with her father's assistance, established the Bell Telephone company. Bell's telephone patent was recognised over competing claims including that of Elisha Gray giving the company a monopoly on the telephone. Bell's priority and honesty in relation to the invention of the telephone have been questioned. Conot claims that a patent examiner was bribed to show Gray's patent caveat to Bell. Bell was later charged with larceny in connection with Antonio Meucci's telephone but the case did not come to trial due to Meucci's death. Bell subsequently withdrew from active involvement in the company that bore his name, turning to non-commercial interests that included founding and becoming president of the National Geographic Society.

structure in order to perform functions. Large technological systems like electric lighting utilities are discussed in Part III.) Using functional systems as an analytical tool, it shows that even simple inventions like Edison's first Phonograph can have a very large numbers of possible solutions, few of which will be successful. This is significant because it refutes a criticism of Edison that he was merely a lucky tinkerer because this low probability of accidental success rules out luck as significant in Edison's huge inventive output.

These concepts are then used to address questions of innovation and novelty. Its most significant conclusion is that a successful innovation is the sum of its failures: the more successful it is, the more it has failed. The consequence of this is that not only should we expect innovations to fail, we should want them to fail. We should also be sceptical of innovations that appear to have been created with few failures since it means either that are likely to fail in the future or that their creation involved little innovation.

To this point, failure has been used in a broad sense. Part I concludes with an examination of a specific type of failure, catastrophic failure, that is a failure in which the artefact is either destroyed or severely damaged, often at considerable economic and human cost. It discusses some examples of catastrophic failures that resulted from altering success criteria so that what previously would have been classed as failures came to be treated as successes but ultimately failed catastrophically. It also concludes that innovation in complex systems is inherently likely to lead to catastrophic failure.

1.2   Part II Edison, Science and Invention

If Edison had a positive attitude to failure, he was even more positive about success. Part II looks at Edison's positive reaction to two notable observations. One of these observations produced a revolutionary invention. The other led to a failure that was so significant it almost ended Edison's plans for his Menlo Park laboratory. The revolutionary invention is the Phonograph. This is approached through a combination of historic analysis of Edison's papers and an experimental replication of his first Phonograph experiments. After his first successful sound recording, Edison wrote in his laboratory notebook that there was "no doubt" he would be able to record and reproduce the human voice "perfectly". Replication of that experiment suggests that this was a remarkably bold claim because what he heard that night would have been unrecognisable as a human voice. Tracing Edison's progress from this first crude experiment to a successful Phonograph instrument 5 months later provides insights into why it was Edison and not someone else who produced the first successful recorded sound.

While the Phonograph was an international success, Edison's public and widely disputed claim to have discovered a new force of nature was not. Like the Phonograph, it began as a remarkable leap from meagre evidence. Seven months before Edison's first sound recording experiment he concluded, based on one observation of an anomalous phenomenon, that he had discovered a new force of nature he called Etheric force. The claim became an early and very public failure for Edison. Yet his failure was not because he was an inventor who did science badly but because, when he moved from inventing to developing a scientific theory, he abandoned the approaches he successfully used as an inventor. Subsequent research revealed that Edison had, indeed, observed something revolutionary, not a new force of nature, but wireless transmissions. Had Edison approached Etheric force differently, he may well have become a pioneer of wireless.

1.3   Part III Edison's World

Part III turns to aspects of the world in which Edison worked. It begins with the American patent system, the system that enabled Edison and other professional inventors to thrive. It also shaped what Edison did. Analysis of patterns evident in Edison's 1086 American utility patents reveals, for example, that he did not invent to with the primary aim of creating large technological systems like electric lighting, but was an opportunist for whom large technological systems presented many opportunities for inventions and patents.

The world of the inventor is, of necessity, one of limited knowledge. Part III examines trial and error, Edison's preferred way of dealing with limited knowledge. When he began developing the carbon microphone there was no existing theory applicable to the vibration sensitivity of carbon. Edison's solution to this lack of theory was to use trial and error, an approach that became so identified with him that it is sometimes called the Edisonian Method. Derided by many (Nicola Tesla called it "inefficient in the extreme"), closer examination reveals that, far from being inefficient and the last resort of the ignorant and uneducated, trial and error is an efficient approach that can be used when no relevant theory exists, including by scientists working at the edge of current theory.

1.4   Part IV Reversing Edison

Previous chapters look at how Edison progressed from the functions he wanted his inventions to perform, to the form the inventions needed to achieve those functions. Part IV asks whether it is possible to reverse this process, that is, use the form of something to determine its function. Specifically, it examines the claim made by philosopher of biology Dan Dennett, that biology is the reverse engineering of natural systems. Drawing on examples from previous chapters and from the techniques of reverse engineering used by engineers, it concludes that we should not be confident that we can identify the functions of something, be it made by humans or nature, from an examination of its form.

1.5 Citing the Thomas A. Edison Papers

Thomas Edison's papers total over three and a half million pages. Selections are published in several forms by the Thomas A. Edison Papers project including an edited and annotated book series, a digital (on line) edition and a microfilm edition. In this book only the first two are cited.

1.6 The Thomas A. Edison Book Edition (TAEB)

Citations designated TAEB refer to documents in the Thomas A. Edison Papers book edition. Currently, the seven published volumes cover the period from Edison's birth in February 1847 to December 1884.[11] Documents in the book edition are numbered consecutively across the volumes and it is the document number, rather than page number, that is cited using TAEB notation. In the citation "TAEB 2:679n5", TAEB indicates the Thomas A. Edison Papers book edition, 2 the volume number, 679 the document number and n5 note 5 to document 679.

1.7 The Thomas A. Edison Papers Digital Edition (TAED)

TAED citations refer to documents in the Thomas A. Edison papers digital edition and use the citation method suggested by the Thomas A. Edison papers editors.[12] For example, in the citation TAED SB1677:126, TAED indicates the Thomas A. Edison

---

[11]Thomas A Edison, *The Making of an Inventor, February 1847–June 1873*, ed. Reese V Jenkins, et al., vol. 1, The Papers of Thomas a Edison (Baltimore: Johns Hopkins University Press, 1989).

From Workshop to Laboratory, June 1873–March 1876, ed. Robert A Rosenberg, et al., vol. 2, The Papers of Thomas a Edison (Baltimore: Johns Hopkins University Press, 1989).

TAEB 3, 3.

The Wizard of Menlo Park, 1878, ed. Paul B Israel, Keith A Nier, and Louis Carlat, vol. 4, The Papers of Thomas a Edison (Baltimore: Johns Hopkins University Press, 1998).

Research to Development at Menlo Park, January 1879–March 1881, ed. Paul B Israel, et al., vol. 5, The Papers of Thomas a Edison (Baltimore: Johns Hopkins University Press, 2004).

Electrifying New York and Abroad, April 1881–March 1883, ed. Paul B Israel, et al., vol. 6, The Papers of Thomas a Edison (Baltimore: Johns Hopkins University Press, 2007).

Losses and Loyalties, April 1883–December 1884, ed. Paul B Israel, et al., vol. 7, The Papers of Thomas a Edison (Baltimore: Johns Hopkins University Press, 2011).

[12]Thomas A. Edison Papers. 2019. "Citing Edison Papers Documents." [web page]. The Thomas Edison Papers, Rutgers, The State University of New Jersey. http://edison.rutgers.edu/citationinst. htm

Papers digital edition, SB1677 is the Folder/Volume ID in the Thomas A. Edison papers digital edition database and 126 the image number in that Folder/Volume. Document images can be accessed through the Thomas A. Edison Papers website using the Folder/Volume ID to locate the folder, then the image number within the folder.[13]

---

[13]2019. "Search Method: Retrieve a Single Document or Folder/Volume." [web page]. The Thomas Edison Papers, Rutgers, The State University of New Jersey. http://edison.rutgers.edu/singldoc.htm

Part I
Edison and Failure

Chapter 2
Success, Failure and Innovation: The Carbon Microphone

2.1 Edison's Dilemma

Thomas Edison was remarkably positive in the face of failure. Towards the end of his life, he related the following incident:

> I never allow myself to become discouraged under any circumstances. I recall that after we had conducted thousands of experiments on a certain project without solving the problem, one of my associates, after we had conducted the crowning experiment and it had proved a failure, expressed discouragement and disgust over our having failed "to find out anything". I cheerily assured him that we had learned something. For we had learned for a certainty that the thing couldn't be done that way, and that we would have to try some other way.[1]

Edison seems to have been fond of thinking of failure in this way because it appears in a number of other anecdotes. After the failure of many experiments on electrical storage batteries he was reported as saying that he did not regard the effort as wasted because "I know several thousand things that won't work" and another time that "I can never find the thing that does the work best until I know everything that don't do it!",[2,3] Clearly, since he was so fond of repeating it, Edison believed that this was how he used failure and did not mention the other many ways he used failure as a tool. It seems that Edison believed the value of failure was for finding "things that won't work".

Some of this chapter appeared previously in Ian Wills, "Instrumentalising Failure: Edison's Invention of the Carbon Microphone," *Annals of Science* 64, no. 3 (2007).

---

[1]Edison, *The Diary and Sundry Observations of Thomas Alva Edison*, 43.

[2]quoted in Frank Lewis Dyer and Thomas Commerford Martin, Edison, His Life and Inventions, (Electronic Text Center, University of Virginia Library: (1998); New York: Harper and Brothers Publishers, 1910), http://etext.lib.virginia.edu/toc/modeng/public/Dye2Edi.html. Online text. 616.

[3]Martin André Rosanoff, "Edison in His Laboratory," *Harpers Magazine* 165, no. September (1932).

© Springer Nature Switzerland AG 2019

I. Wills, *Thomas Edison: Success and Innovation through Failure*, Studies in History and Philosophy of Science 52, https://doi.org/10.1007/978-3-030-29940-8_2

The implication that he succeeded by conducting a large number of experiments that failed until he eventually stumbled on one that worked cannot explain Edison's many successes. The fundamental problem with trying to succeed by building a long list of things that do not work is that this knowledge alone tells nothing of what might work. Edison clearly knew how to make things that worked and certainly revelled in success, so how did he use these many failures to arrive at eventual success? The first step in answering that is to recognise that Edison did not regard failure as a negative to avoid but as something actively pursued. It was a tool for moving towards success.

To see how Edison used failure as a tool, we will follow development of the carbon microphone, crucial to making the telephone a technical and commercial success and a technology that was central to the telephone for the next century. Anyone who has used a telephone with a rotary dial will almost certainly have spoken through a version of Edison's carbon microphone.

2.2 Edison and the Telephone

Although the name of Alexander Graham Bell is most often associated with the invention of the telephone, Edison played a significant part in making it viable through his invention of the carbon microphone. He was not the only claimant to a patent for this critical device and became involved in a 15-year legal battle with eight others, including Alexander Graham Bell, Elisha Gray (founder of the Western Electric Company) and Emile Berliner, inventor of the Gramophone.[4,5,6] These competing claims (patent interference) ended in 1892 when Edison was issued with two patents for the carbon microphone. By that time the Bell Telephone Company had purchased the rights to both Edison's and Berliner's designs.[7] Significantly, the litigation did not include Italian born American, Antonio Meucci, recognised by the United States Congress in 2002 as the inventor of the telephone.[8,9]

---

[4]TAED TI1:2.

[5]Elisha Gray (1835–1901) American inventor entrepreneur. Gray was a prolific Chicago based inventor of electrical devices and partner in the firm that was later to become Western Electric.

[6]Emile Berliner (1851–1929) German American inventor. Berliner emigrated from Germany to America and, like Edison and Gray, was a professional inventor-entrepreneur. His best known invention is the gramophone, a device similar to Edison's phonograph. The gramophone employed flat disks that eventually displaced Edison's cylinders.

[7]Thomas A Edison. Speaking-Telegraph [1]. US Patent 474,230, filed 27 April, 1877, and issued 3 May, 1892; Speaking-Telegraph [2]. US Patent 474,231, filed 20 July, 1877, and issued 3 May, 1892.

[8]*Resolution 269*. Bell was charged with larceny in connection with the theft of Antonio Meucci's telephone model but did not go to trial due to Meucci's death.

[9]Antonio Meucci (1808–1889) Italian American inventor and manufacturer. Meucci demonstrated a telephone in New York in 1860.

2.3 Challenges to Western Union

Edison's invention of the carbon microphone was the consequence of the battle for control of the telegraph industry fought, in part, through patents. In the 1870s, the American telegraph industry was dominated by Western Union, which at one point transmitted 90% of all telegraph messages in the United States.[10,11] Alexander Graham Bell's telephone patent and Elisha Gray's harmonic telegraph patents represented a threat to this dominance.[12,13] Both technologies had the potential to dramatically reduce the cost of sending telegraph messages for the telegraph company that controlled them, Bell's by sending them as spoken messages rather than by Morse code and Gray's by sending many Morse code messages simultaneously down the same wire.

Western Union countered these threats in a number of ways, one of which was to commission Edison to invent devices to circumvent Bell's and Gray's patents. Edison's telegraphy inventions, notably the duplex and quadruplex telegraphs (which enabled two and four messages respectively to be sent simultaneously on the same wire), and his improvements to the printing telegraph (stock ticker) had significant commercial impact. This ability to produce successful inventions, seemingly on demand, brought Edison to the attention of the men behind the telegraph companies, notably Cornelius Vanderbilt, the largest stockholder in Western Union, and its president, William Orton.[14,15]

---

[10]Robert E Conot, *A Streak of Luck*, 1st ed. (New York: Seaview Books, 1979), 37.

[11]The Western Union Telegraph Company was formed in 1856 following a number of mergers of smaller telegraph companies. At one time Western Union transmitted 90% of telegraph messages, entering the money transfer business in 1871 and, in 1878, telephone business using Edison's telephone inventions. When Bell Telephone challenged the legality of the latter venture, Western Union's sold its telephone interests to Bell. The telephone and later technologies led to a decline in Western Union's business to the point that currently only the money transfer business remains. Western Union delivered its last telegram on Friday 27 January 2006.

[12]Alexander Graham Bell. Improvement in Telegraphy. US Patent 174,465, filed 14 February 1876, and issued 7 March 1876.

[13]Elisha Gray. Improvement in Electric Telegraphs for Transmitting Musical Tones. US Patent 166,095, filed 19 January 1875, and issued 27 July 1875; Improvement in Transmitters for Electro-Harmonic Telegraphs. US Patent 165,728, filed 28 June 1875, and issued 20 July 1875; Improvement in Receivers for Electro-Harmonic Telegraph. US Patent 166,094, filed 28 June 1875, and issued 27 July 1875; Improvement in Local-Circuit Breakers for Electro-Harmonic Telegraphs. US Patent 194,671, filed 15 February 1875, and issued 28 August 1877.

[14]Cornelius Vanderbilt (1794–1877) American Civil War Commodore, shipping and railway magnate and one of the richest men in America. Patriarch of the Vanderbilt financial empire, Cornelius built his wealth first in shipping then railroads. Vanderbilt was also a major stockholder in Western Union, making him a competitor to Jay Gould in both railroads and telegraph.

[15]William Orton (1826–1878) American industrialist. Orton became President of Western Union in 1867 and continued in the position until his death. It was Orton who arranged for Western Union to finance Edison's plans to build his research laboratory at Menlo Park.

When Edison patented his quadruplex telegraph system, he initially offered it to his former employer, Western Union. Since the quadruplex enabled four messages to be sent simultaneously on one wire, it had the potential to increase the revenue producing capacity of Western Union's network fourfold with little additional capital cost. At the end of 1874, negotiations with Western Union had dragged on for months without resolution so Edison, under pressure from his creditors, licensed the quadruplex to Western Union's rival, Jay Gould's Atlantic and Pacific Telegraph Company.[16] Gould, who had ruthlessly built a railway empire, intended to repeat his success with the telegraph.[17]

During his negotiations with Western Union, Edison supplied Western Union with several prototypes of the quadruplex. Western Union's president, William Orton, dealt with Edison's defection to Gould by ignoring his patents and had the prototypes copied. This was something Gould could not ignore. In the court battle that followed, Edison appeared as a witness for Gould's Atlantic and Pacific Telegraph Company but, having resolved his differences with Western Union, Edison spent much of his time during the hearings in Western Union's offices.

The quadruplex case was a salutary lesson for Orton, who determined to keep Edison in the Western Union camp rather than risk his inventive talents being used by competitors. Edison also drew several lessons from the quadruplex case. One was to be wary of becoming "a tool of Wall Street" although he was obliged to draw on Wall Street capital to exploit his inventions until he became independently wealthy. In 1893, after losing control of his electrical companies, Edison determined to be his own entrepreneur in future. The other lesson was to record the development of his inventions in laboratory notebooks which could later be tendered as evidence in the inevitable event of his patents being contested. Those laboratory notebooks provide the basis for this study of Edison at work.

When Western Union could not overcome the Atlantic and Pacific Telegraph Company by other methods, it neutralised the threat by forming a price-setting cartel with Gould, later buying Gould's company. The move was typical of trusts in post-Civil War America, Western Union's telegraph system and Jay Gould's railways being just two. Edison came to understand the value of his inventions to reinforce or, in the hands of others, diminish Western Union's near monopoly of the telegraph. The potential of his quadruplex to significantly reduce costs could have made the Gould's Atlantic and Pacific Telegraph Company a serous rival to Western Union.

---

[16]Conot, *A Streak of Luck*, 66.

[17]Jay Gould (1836–1892) American financier. Gould seemed to set no limits on what he might do to achieve his financial ends. When he invested heavily in railroads Gould resorted sometimes to devious means to take over smaller railroads.

2.4   Edison Starts Work on the Telephone

Edison started work on circumventing Bell's telephone patent in mid-1875 with a number of disadvantages. Unlike Bell, who as a teacher of the deaf with a thorough understanding of the mechanisms of speech and hearing, Edison knew little of this and was partially deaf himself. Edison also began with only two exemplars, both of them of limited value. The first was Bell's telephone, which he could not use because of Bell's patent while the second, the German Philip Reis's *telephon*, transmitted only musical tones and not speech.[18,19]

Figure 2.1 shows the final form of Edison's carbon microphone, the design that Western Union put into commercial production in mid-1878. In the carbon microphone, sounds enter the horn shaped mouthpiece (M) causing the diaphragm (D) to vibrate. This vibration is transmitted via an aluminium button (A) to carbon particles (C) causing their electrical resistance to vary in sympathy with the sound. The screw S can be used to compress the carbon particles (Edison used lamp black, purified soot from oil lamps) which are confined in a small space but able to deform under pressure. When pressure caused by the entering sound is applied to the lamp black, this deformation changes the contact area between adjacent particles, and hence the electrical resistance. For a constant applied voltage, Ohm's law indicates that the electrical current passing through the carbon also varies in sympathy with the sound.

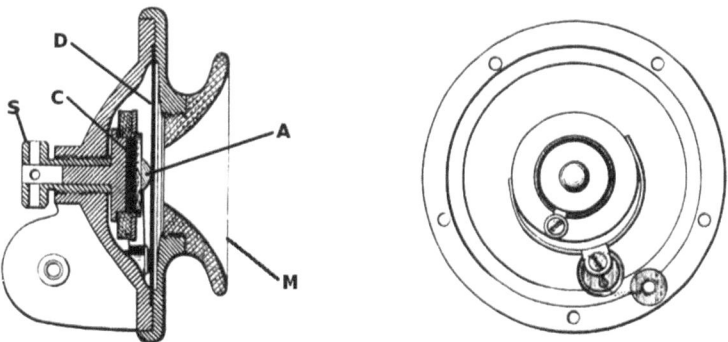

Fig. 2.1   Edison's telephone transmitter (notation added). (TAED TI2:490)

---

[18]V Legat 1862. "Reproducing Sounds on Extra Galvanic Way." http://edison.rutgers.edu/singldoc.htm as TI2:456–458.

[19]Philip Reis (1834–1874) German inventor, and teacher of mathematics and physics. Reis built and demonstrated a device for transmitting sounds by electricity. It was Reis who coined the name telephone, ("Telephon" in German). Reis's telephone apparatus was demonstrated in New York in 1872 and described in Baile's, *Wonders of Electricity* a book read by Edison and Bell. Edison based his first telephone sketches on Reis's instrument.

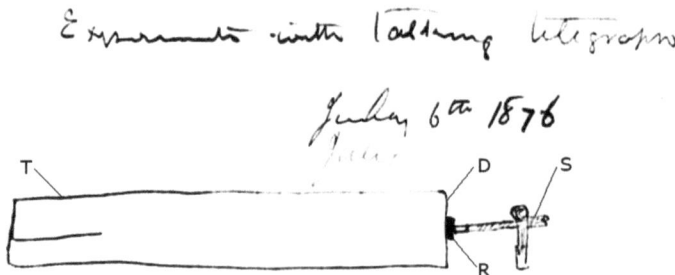

Fig. 2.2 Detail of Edison's sketch of 6 July 1876. The handwritten text reads "experiments with talking telegraph July 6th 1876". The electrolyte saturated felt variable resistance is the black rectangle (R). A screw (S) adjusts the pressure of the felt against the diaphragm (D) at the end of the speaking tube (T). (TAED TI2:34)

Figure 2.2 is one of Edison's earliest dated telephone sketches. It shows a speaking tube (T) at the end of which is a diaphragm (D). The variable resistance (R) (a piece of felt saturated with electrolyte) is pressed against the diaphragm by a screw (S). Apart from the material used as a resistance and the shape of the speaking tube, the device uses the same operating principle to the production version in Fig. 2.1. This similarity appears to support the popular belief that invention consists of an initial flash of inspiration in which the inventor conceives the successful solution followed by a period in which the remaining details are worked out. Following Edison through the development of the carbon microphone will show that this simplistic view is far from the inventor's experience because the path from initial concept to final solution was far from linear, with many false starts, dead ends and back tracking, and many, many failures.

2.5   Greenwich, England, 1873

Edison's carbon microphone had its origins in 1873, on the other side of the Atlantic, and with an embarrassing failure. In April 1873, 26-year-old Edison travelled to England to promote his automatic telegraph system, which was intended to increase the speed of telegraph transmissions by sending messages at a much higher speeds than they could be tapped out manually in Morse code or transcribed by the receiving operator. In use, operators at the sending end of the telegraph line recorded telegraph messages on perforated paper tape using conventional telegraph keys. This tape was then fed through a machine that transmitted the resulting code at high speed to the receiving end, where another machine, the recording telegraph, recorded the message as dots and dashes on chemically treated paper tape, the code being subsequently transcribed into text by telegraph operators.

Edison's trip began with some promise when he demonstrated the recording telegraph to the British Post Office, achieving speeds of 500 words per minute over short distances. This was slower than the 1000–1500 words per minute Edison claimed for his invention, but higher than achieved by the rival Wheatstone system, invented by the physicist Charles Wheatstone and used by the Post Office since the 1860s.[20,21]

To demonstrate the suitability of his recording telegraph for transatlantic transmissions, a further test was arranged using a cable at the Greenwich works of the Telegraph Construction and Maintenance Company. At the time, this company manufactured and laid most of the world's oceanic cables and had available 3500 km long cable awaiting laying between Europe and Brazil, the cable being stored in huge coils prior to loading onto a cable laying ship.[22] Despite never having worked with oceanic cables directly, Edison seems to have been confident of the suitability of his invention for this purpose.

The demonstration was anything but a success. As Edison told the story, the first dot he transmitted was printed not as a single dot on the recording tape but as "17 foot" (5 m) long dash. Despite working with his equipment for a fortnight, the highest speed Edison achieved was two words per minute, only a seventh of the guaranteed speed of the cable and far less than the 500 words per minute he had demonstrated earlier. Crucially, it was also less than the 10–17 words per minute achieved manually on transatlantic telegraph transmissions.[23]

2.6 Exploring Induction

The trip across the Atlantic made several negative impressions on the young Edison. The Atlantic crossing was so unpleasant that Edison named his shop, the *SS Java*, the "Jumping Java", the crossing souring Edison against overseas travel for life. (The only other overseas journey Edison made was a triumphal tour of Europe in 1889.)[24] Further, although the British Post Office reimbursed the cost of the demonstration, it did not buy Edison's invention, making the trip a commercial failure. Finally, the Greenwich demonstration was an embarrassing personal failure for a young man who prided himself on his knowledge of telegraphic technology.

In the circumstances, it would have been understandable had Edison to put the whole experience behind him and to move onto other projects. That was not Edison's

---

[20]TAEB 2:591.

[21]Charles Wheatstone (1802–1875) English physicist and inventor. Wheatstone was prominent in the telegraph field and is best known for his invention of the Wheatstone bridge, a device for accurately measuring resistance. Apart from his inventions in telegraphy, Wheatstone also invented a musical instrument, the concertina.

[22]Edison, *TAEB* 2, 2, 5.

[23]Frank Lewis Dyer and Thomas Commerford Martin, Edison, His Life and Inventions, (Electronic Text Center, University of Virginia Library: (1998); New York: Harper and Brothers Publishers, 1910), http://etext.lib.virginia.edu/toc/modeng/public/Dye1Edi.html. Online text. 151.

[24]Conot, *A Streak of Luck*, 76.

way. Instead, as on many other occasions, he seized the Greenwich failure as a challenge. He may have been a puzzled man in Greenwich, but soon realised that the cause of his problem was electrical induction caused by the coiling of such a long cable. That he should have overlooked induction is itself puzzling for he claimed to have read and thoroughly digested Faraday's *Experimental Researches in Electricity* as early as 1868.[25] In Greenwich, Edison seems not to have connected the results of Faraday's experiments on small induction coils to the 3500 km coil of telegraph cable. Faraday had found that when a current passes through a wire it produces a magnetic field around the wire. Coiling the wire magnifies the magnetic field. Although the cable would have exhibited little inductance laid in a straight line under the ocean, when closely wound in its wells at Greenwich it became series of massive inductors. When Edison pressed his telegraph key, the coil stored the electrical energy as a large magnetic field then, when he released the key, the field collapsed, releasing electrical energy to produce the "17 foot dash".

Edison's realisation that his embarrassing failure at Greenwich was caused by induction piqued his interest to the point that he put aside inventing for much of the next year in order to explore induction phenomena, his patent output dropping to a seventh of what it had been the previous year. Edison continued his habit of experimenting night and day, as he had before he left for England, but focused on understanding induction rather than producing patentable inventions. His failure at Greenwich may have been a disappointment, but it highlighted oceanic cables as an inventive challenge, and Edison knew that inventive challenges were the catalyst of commercial inventions.

Once Edison had identified induction in the coiled cable as the cause of his Greenwich failure, he set about devising means of simulating the behaviour of long oceanic cables while avoiding induction effects. For this, he needed test devices that could simulate the capacitance, low inductance, and high resistance of the oceanic cables. Since these were not commercially available, Edison invented a high resistance rheostat (adjustable resistor) consisting of "heavy glass tubes filled tightly with flour of gas retort carbon", the tubes being "1/16 or 1/32 bore" (1.6 or 0.8 mm).[26] This was to be built in two versions that could simulate resistances of 1,000,000 ohms in steps of 50,000 ohms and 100,000 ohms in steps of 5000 ohms. Together these ingenious devices would have given Edison any resistance from 5000 ohms to 1,100,000 ohms in 5000 ohm steps, all in a very compact form.

Edison ordered ten of these but only one was finished (Fig. 2.3) because when tested, it failed to work as expected. Instead of the stable resistances he needed to simulate the cable, Edison "found that the resistance of carbon varied with every noise, jar or sound, and [the rheostats] were too unreliable where a definite resistance was required". He also noted, "The pressure of the electrodes in contact with the

---

[25]TAEB 1:7n3.
[26]TAEB 1:345

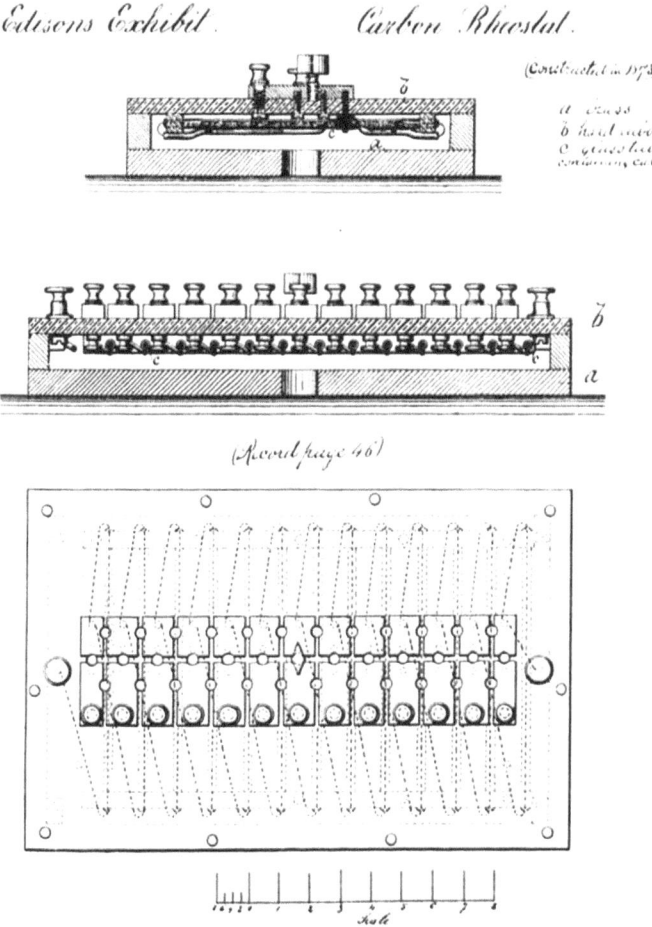

Fig. 2.3 Edison's 1873 carbon rheostat from his Telephone Interferences testimony. (TAED TI2:466)

carbon varied the resistance".[27] Edison abandoned the carbon rheostat for wire wound resistances but the recollection of the observed vibration sensitivity of carbon was to prove of value later.

One of the few instances of Edison's patent-directed work during this period dealt with diplex telegraphy.[28] The diplex enabled two telegraph messages to be sent in the same direction on the same wire and exploited reverse currents, an induction effect that occurs when the field in an electromagnet (in this case in a relay) collapses

---

[27]TAED TI1:41
[28]TAEB 2:348

following breaking of the circuit. Reverse currents in the relay sending one message caused unwanted movement in the relay sending the other message. They were, to use a term that Edison coined, a bug.[29]

To deal with the bug, Edison devised what he termed a "bug trap", in this instance another relay, adjusted to act slowly, so the reverse current did not interfere with the signal. Bug traps were a strategy that Edison used when he could not eliminate the cause of a problem, their objective instead being to render the effect of the bug insignificant.[30] The diplex telegraph system was to become the basis of the quadruplex telegraph, an even more ingenious and valuable invention that combined two diplexes, enabling four messages to be sent simultaneously on the same cable, two in each direction.[31]

2.7 Bell's Telephone

Bell's telephone was initially seen by Western Union, Edison, Gray and other telegraph experts as just another means for reducing the cost of sending telegraph messages.[32] While Edison's quadruplex and Gray's acoustic telegraph sent multiple messages using Morse code over the same telegraph wire, Bell's telephone would permit transmission at speaking speed. Instead of requiring skilled telegraphers, messages could be sent over telephone by cheaper untrained operators. Used in this way, a telephone operator would read the message over a telephone line to another operator at the receiving end, who would transcribe it to written text for delivery to the recipient. Except for transmission by speech, this was the process used for sending conventional telegrams. Neither Edison nor Orton imagined that the telephone would be used for direct person-to-person verbal communication.

Bell's invention was promoted by his father-in-law, Gardiner Hubbard, a prominent Boston lawyer and financier.[33] Hubbard approached Orton with an offer of the patent rights to the telephone, then in its early stages of development. Orton was

---

[29]Edison used the term bug so early and so frequently that he seems to have coined it. This patent, from August 1873, predates the earliest citation (1889, also from Edison) in the Oxford English Dictionary. Oxford English Dictionary, "Bug, N.2."

[30]Edison, *TAEB 2*, 2, 4.

[31]Ibid., 21.

[32]David A Hounshell, "Elisha Gray and the Telephone: On the Disadvantages of Being an Expert," *Technology and Culture* 16, no. 2 (1975).

[33]Gardiner Hubbard (1822–1897) Lawyer and financier. A member of a prominent Boston family, Hubbard engaged Alexander Graham Bell to teach his daughter, Mabel Gardiner Hubbard, who had become deaf as a result of contracting scarlet fever as a child. (Bell later married Mabel.) Through this contact, Hubbard encouraged Bell to develop telephone and with him established Bell Telephone company with Hubbard as its first president. Conot claims Hubbard was involved in dishonest activities in support of Bell including bribing patent office officials. Hubbard became a major stockholder in the Edison Speaking Phonograph Company.

reluctant to purchase the rights, not only because of the cost of acquiring another infant technology, but because Hubbard had been a vigorous opponent of Western Union's telegraph monopoly. As he had done with Gray's acoustic telegraph, Orton engaged Edison to invent a system that would circumvent Bell's patent. If Edison could invent alternatives to Bell's system, the resulting patents would benefit Western Union if it adopted telephone technology, reduce Bell's lead and inhibit other potential competitors. Orton's decision was vindicated later when Western Union sold its telephone business and the patent rights to Edison's carbon microphone to the Bell Telephone Company, at a profit. Notwithstanding Hubbard's objection to Western Union's monopolistic approach to the telegraph, possession of Edison's enabling patent[34] for the carbon microphone helped Hubbard, as the first president of Bell Telephone, to establish it as an even greater monopoly that endured for a century.

Edison claimed in his legal testimony in the carbon microphone patent interference that, soon after his meeting with Orton in mid-1875, Orton had provided him with a translation of a German report describing a promising device. This report, by V Legat, inspector of the Royal Prussian Telegraph, described an 1861 demonstration by the German Philip Reis of a device Reis named the *telephon*. In the report, Legat observed that with Reis's apparatus "[chords], melodies, &c — are transferred with astonishing correctness" but "the vocal seems more or less indistinct".[35] At another point in his testimony, Edison said he had seen Reis's apparatus demonstrated by Vander Weyde in New York in 1872, and to have read a description of it in Baile's *Wonders of Electricity*.[36] (Bell also mentions seeing the Vander Weyde demonstration and reading Baile.) Edison commented that the Legat report contained nothing more than he had previously read about Reis's device.[37,38]

Figure 2.4 is a drawing of Reis's apparatus from the Legat report. The upper part shows the transmitter consisting of a megaphone style mouthpiece, the small end of which contains a diaphragm. Movement of the diaphragm in response to the entering sound opens and closes the electrical circuit through a sensitive switch. The lower part of the drawing shows the receiver in which the resulting electrical current passes through an electromagnet, which in turn causes a reed to vibrate, emitting sound.

At the time (1875), Bell was using an instrument, known as the Gallows Telephone, that functioned both as transmitter and receiver (Fig. 2.5). This consisted of a small permanent magnet attached to a diaphragm (M) that vibrated close to an

---

[34]See Chap. 9 for a discussion of the significance of enabling patents.

[35]Legat, "Reproducing Sounds on Extra Galvanic Way"; ibid.

[36]J Baile, *Wonders of Electricity*, trans. John W Armstrong (New York: Scribner Armstrong and Co, 1872).

[37]TAED TI1:21.

[38]Peter Henri Vander Weyde (1813–1895) Dutch born American inventor, scientist and author. He took some of the earliest Daguerreotype photographs in America and became an authority on electrical technologies. Vander Weyde's demonstration of Reis's telephon in New York in 1873 was seen by both Bell and Edison.

Fig. 2.4 Reis telephon. The
upper part is the transmitter
in which sounds cause a
diaphragm to vibrate in turn
opening and closing an
electrical circuit through a
sensitive switch. (TAED
TI2:458)

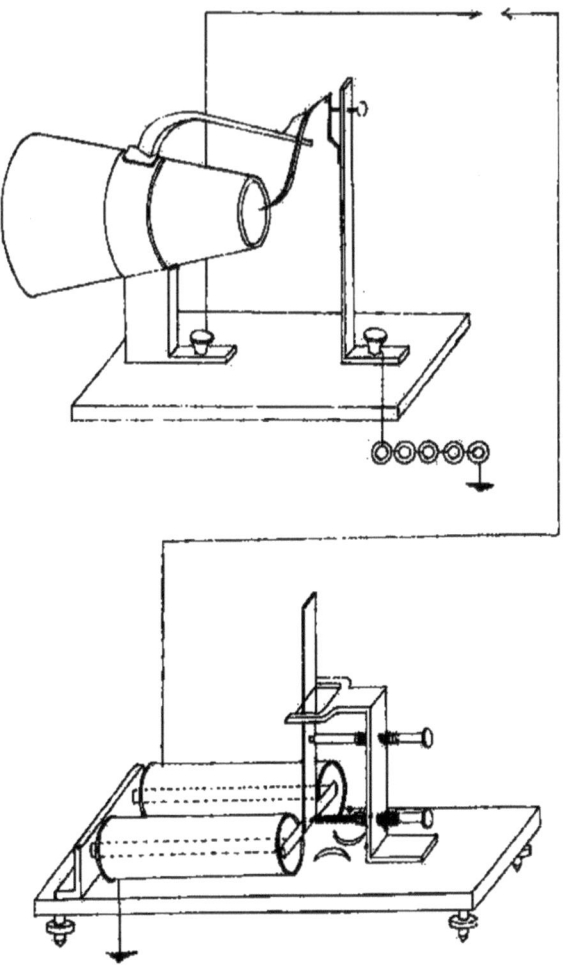

induction coil (C) supported in a gallows-like structure (U). As a transmitter, the
vibrating diaphragm-magnet assembly created a small current in the induction coil.
When used as a receiver, the current from the transmitter passed through the
induction coil causing the magnet-diaphragm to vibrate producing a weak sound.
This system operated without batteries but produced very little current.

Edison knew from his extensive telegraph work, that although Bell's telephone
might transmit speech, its signal was so weak that it could transmit it only over
distances of a few metres. The Reis device, on the other hand, used an impressed
current like the telegraph and potentially had a much greater geographic range
because it could use much larger currents from batteries. However, it failed to
transmit speech because it used a sensitive switch (Fig. 2.4) to break the circuit in
sympathy with the sound vibrations, an arrangement that Edison recognised could
transmit only pure tones of constant intensity and would not reproduce the varying

Fig. 2.5 Bell's first "gallows" telephone transmitter-receiver from about July 1875. (Michael E Gorman. 1994. "Alexander Graham Bell's Path to the Telephone." [Web site], Last Modified 07-Dec-1994. http://www2.iath.virginia.edu/albell/fgt.2.html)

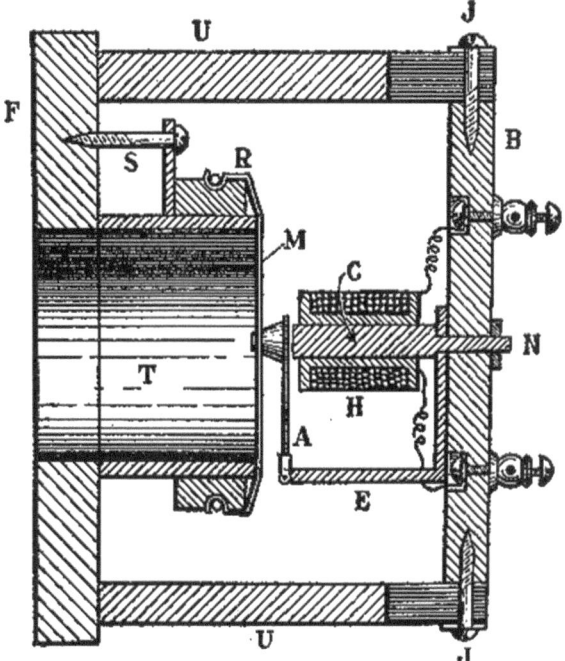

acoustic volume of loud and soft speech nor harmonics.[39] From these shortcomings, Edison decided that the key to successfully competing with Bell was to create a device that could transmit articulate speech as Bell's did, but also transmit it over hundreds of kilometres as the telegraph did.

Edison therefore started work on the telephone with Bell's and Reis's devices as exemplars. While we might expect that Edison would have exploited the successful parts of each, he appears to have started by identifying where each failed.

In his legal testimony in the telephone interference case, Edison claimed that he sketched (but did not date) three alternative telephone designs in July 1875 (Fig. 2.6). Edison knew of Helmholtz's experiments in reproducing vowel sounds and his top sketch (which appears to have been the first one drawn) draws on Helmholtz's apparatus (Fig. 2.7). It shows a cylindrical speaking tube connected, via a tuning fork, to a variable resistor consisting of an electrode making contact with mercury and the notation "mercury like Helmholtz". For the middle sketch, Edison drew directly on Reis's design but replaced the sensitive switch with a liquid variable resistor similar to one in a telegraph relay he had patented previously.[40] He completed the circuit with a receiver and battery. The bottom sketch shows a similar speaking tube and diaphragm arrangement but the diaphragm (labelled "parchment")

---

[39]TAED TI1:23.

[40]Thomas A Edison. Relay Magnets. US Patent 141,777, filed 13 March, 1873, and issued 12 August, 1873.

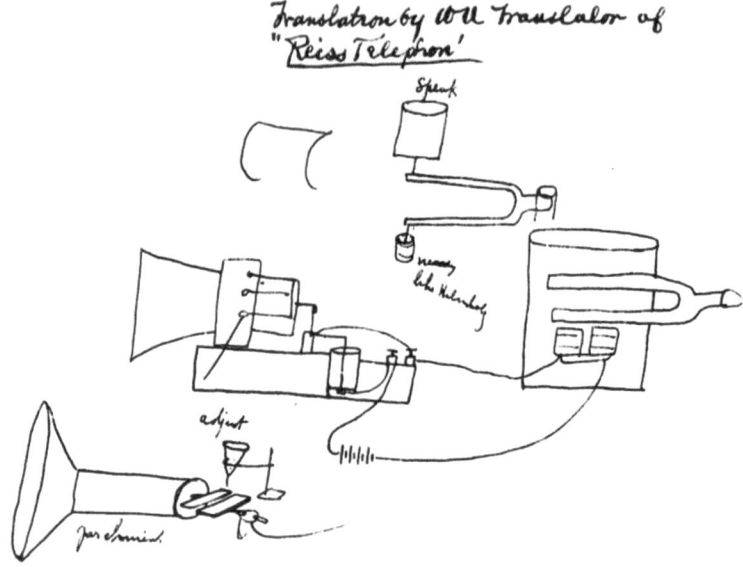

Fig. 2.6 Edison's first telephone sketches, July 1875. The notes read: "Translation by WU [Western Union] Translator of" "Reis telephon""Speak" "mercury like Helmholtz" "adjust" and "parchment". (TAED TI2:455)

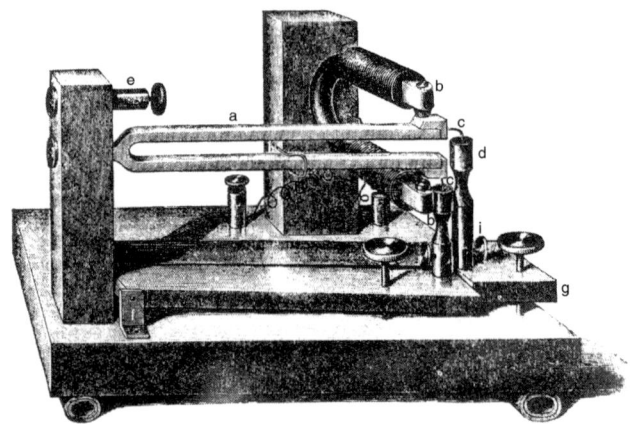

Fig. 2.7 Apparatus used by Helmholtz to produce sounds. (TAEB 2:599n5)

is connected to two knife-edges, separated by a small gap. Electrolyte drips onto this gap from a funnel above it, Edison conceiving this as a way of varying the resistance between the knife-edges in response to the small movements of the diaphragm. The note above the funnel marked "adjust" indicates that the rate of dripping was to be controlled as was the gap between the knife-edges since a screw is indicated connected to the non-vibrating one.

Edison's first sketches contain few words and show him "thinking on paper", recording telephone designs he had pictured mentally, testing ideas and then sketching new arrangements that addressed the failures he had identified in the previous sketches. Edison described the sketches as "rough ideas of how to carry out that which was necessary in my mind, to turn the Reiss transmitter into an articulating transmitter. They were notes for future use in experimentation".[41] This is an apt description of most of Edison's sketches, which generally contain little detail or notes and seem to have been executed quickly, perhaps to illustrate discussions with associates. These sketches suggest that Edison's primary concern was to record general concepts and manipulate ideas visually. Only one of these first three sketches shows a receiver, suggesting that Edison was concentrating on the transmitter as the crucial part of the telephone and had identified the variable resistor as the key to giving the Reis apparatus the ability to transmit articulate speech. Edison described the sketches as being devices for "varying the resistance of a circuit, containing a constant current, by the energy of the voice".[42] In all three sketches, Edison replaced Reis's switch with a variable resistor, a key feature of his final, successful, telephone transmitter. Before he arrived at his final design, there were many other possibilities to be conceived, tested and eliminated.

Edison claimed he drew these undated sketches in July 1875, but there is a gap of almost a year before his first dated records of work on the telephone. Edison's explanation for the lack of records was that work on the telephone during 1875–76 was recorded by an associate, James Adams in a personal notebook that was lost after Adams's death.[43,44] In his testimony, Edison claimed to have built an instrument in October or November 1875 "to ascertain the availability of exceedingly slight motions of immersed electrodes to increase or decrease the resistance of the circuit".[45] He said that some of the instruments built at the time were capable of being used for articulate speech but admitted that they had not been tried for the

---

[41]TAED TI1:22.

[42]TAED TI1:22.

[43]TAED TI1:36.

[44]James Adams (? - 1879) was an English born seaman who joined Edison in Newark as night watchman but soon became his associate in inventing, undertaking much of the early development of the telephone. Adams travelled to Europe to work on telephone installations and died in Paris, possibly of tuberculosis.

[45]TAED TI1:28.

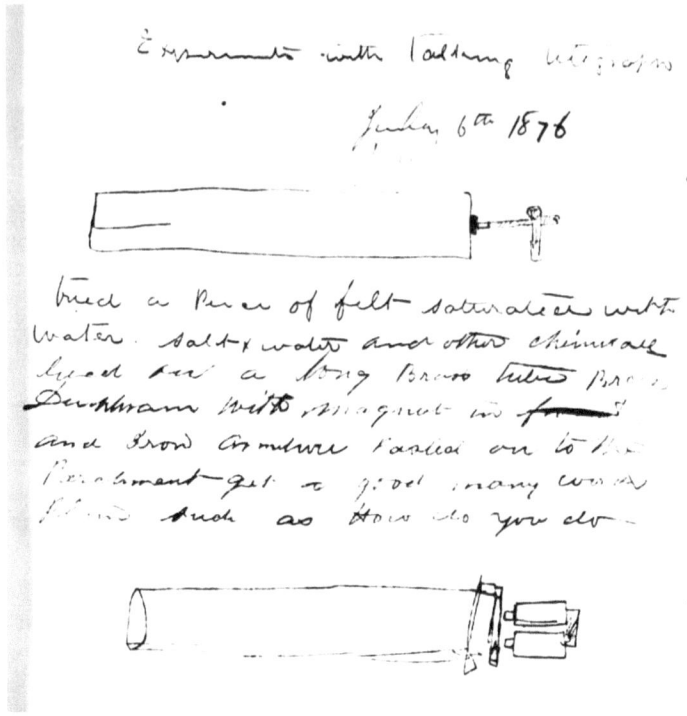

Fig. 2.8 Edison's telephone sketches 6 July 1876. (TAED TI2:34)

purpose.[46] Francis Jehl,[47] another of Edison's associates, later claimed that Edison had built the bottom design in Fig. 2.6 at this time but found it ineffective.[48]

Whether the experiments Edison claimed that he and Adams conducted towards the end of 1875 took place is questionable. The sketches he made in July 1876 (for example, Fig. 2.8), appear very preliminary rather than developments of designs he had built earlier. Given Edison's way of working, it is also unlikely that he would have built instruments and not tried them with articulate speech, the object of his work. Edison made these claims in testimony in a dispute over patent priority so it is possible he was stretching the truth to support his case. Likewise Jehl's account, although presented as first hand, cannot be as Jehl did join Edison until the end of 1879.

---

[46]TAED TI1:20–21, 28.

[47]Francis Jehl (1860–1941) joined Edison at Menlo Park in November 1878 aged 19. He had studied at Cooper Union, been a law clerk and worked in the machine shops of Western Union. Working with Edison, he played a prominent role in the development of electric lighting. In the 1930s, Jehl became the first custodian of Henry Ford's reconstruction of Edison's Menlo Park complex at Dearborn, Michigan.

[48]Francis Jehl, *Menlo Park Reminiscences: Written in Edison's Restored Menlo Park Laboratory*, 3 vols., vol. 1 (Dearborn, Michigan: Edison Institute, 1937), 109.

2.8   Edison's 1876 Telephone Experiments

Bell applied for a patent on the principle of the telephone in February 1876.[49] At the time, he did not have a satisfactory device to demonstrate it but by June 1876, Bell exhibited a working telephone at the Philadelphia Centennial Exhibition. Given the lack of documentary evidence of Edison working on the telephone before mid-1876, it seems more likely that it was Bell's Centennial Exhibition demonstration (or perhaps more significantly, Western Union's concern about the threat it posed), that prompted Edison to start work on the telephone in earnest.

While we might be sceptical about what Edison claimed he did before mid-1876, he recorded experiments on the telephone a few weeks after Bell's demonstration. Edison's first dated records of his experiments on the telephone are sketches from 6 July 1876 (Fig. 2.8). The sketches show a transmitter and receiver (upper and lower respectively in Fig. 2.8), Edison recording that with these instruments he could get "a good many words Plain such as How do you do".[50] Both the transmitter and receiver bear a strong resemblance to the acoustic telegraph (Fig. 2.9) that Edison had been working on in the latter part of 1875. This not surprising, since the acoustic telegraph shared common elements with the telephone including the diaphragm driven receiver.

Edison often approached an invention problem by building and testing many alternative solutions. In July 1876, he did this when he appeared to have cannibalised his acoustic telegraph apparatus for the telephone. During July and August, Edison tested designs based on this arrangement using a variety of operating principles as well as trying different materials and thicknesses for the diaphragm. On 12 July, he sketched a transmitter based on Reis's concept but using multiple switches attached by strings to various parts of the diaphragm, each string and switch intended to detect to a different frequency (Fig. 2.10). The same page of sketches shows other transmitters using multiple tuning forks and strings. These designs failed to transmit articulate speech, but indicate that Edison was not fully committed to the variable resistor approach nor, indeed, to any aspect of the concepts in his initial undated sketch. On 26 July, he sketched more transmitter designs in a variety of arrangements including one with four speaking tubes. These designs are more complicated than that shown in Fig. 2.2 and each appears to have arisen either out of Edison's efforts to deal with failures of previous test models or out of attempts to understand how such a device might function. The multiple string design (Fig. 2.11), for example, came from Edison's hypothesis that high notes were produced in the middle of the diaphragm and low notes at the edges. Tests showed this not to be the case, eliminating the design and the hypothesis. The test provided Edison with a negative example but, more importantly, added to his understanding of the way in which diaphragms function. At this stage, Edison's understanding of the mechanics of the diaphragm was mistaken. Instead of using a human analog, the tympanic

---

[49]Bell. Improvement in Telegraphy.
[50]TAED TI2:34.

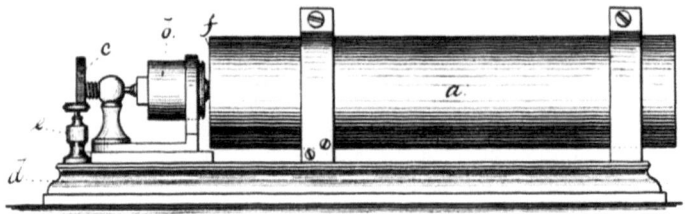

Fig. 2.9 Acoustic telegraph receiver of November 1875. The cylindrical tube and adjustment screw suggest a close resemblance to Edison's telephone sketches of July 1876 (Fig. 2.2). (TAED TI2:463)

Fig. 2.10 Edison's telephone sketches 12 July 1876. (TAED TI2:24)

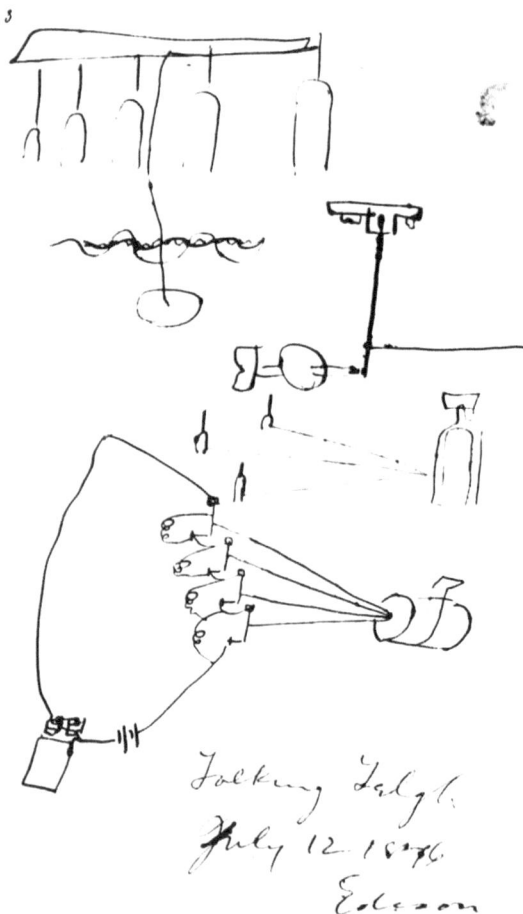

Fig. 2.11 Multiple string
telephone 26 July 1876.
Notation is "This may
work" and "speak here".
(TAED TI2:36)

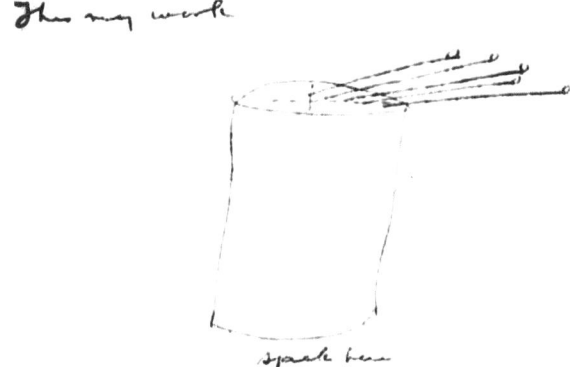

membrane of the ear as Bell, a teacher of the deaf would have done, Edison's limited
understanding of the physiology of speech and hearing led him to model the
microphone diaphragm on the skin of a drum.

At this point Edison did not know that the microphone diaphragm resonates in
sympathy with the incoming sound across its whole surface not at different frequen-
cies in different parts. Problems arising from Edison's misunderstanding of the
mechanics of sound persisted for most of the development of the carbon microphone. Another misunderstanding was Edison's belief that he required a relatively
flexible way of coupling the incoming sound to the electrical components. Similarly,
his modelling of the diaphragm on a drum led him to use relatively flexible materials
like parchment, whereas his final design (Fig. 2.1) worked because it used a
relatively rigid metal diaphragm and metal coupling button.

The most satisfactory configuration from these early efforts used a single diaphragm attached to a platinum wire dipping into an electrolyte.[51] In this arrangement, the movement of the wire varied the contact area with the electrolyte and
hence the resistance, making it very similar to the arrangement in Edison's middle
telephone sketch in Fig. 2.6. This may have confirmed Edison's original idea, but his
other sketches show him trying alternatives and seeking to understand the principles
behind the behaviour of components. Indeed, not only at this early stage but
throughout the next 2 years, Edison periodically returned to earlier designs including
Reis's, or tried completely new concepts. Jehl lists 27 other designs of telephone
transmitter tried by Edison in addition to the final lamp black design.[52] These
designs include several using capacitances; one using a voltaic pile (battery); several
variations on contact switches (like Reis's); several using carbon impregnated silk
disks as the variable resistances; two with the armature of an electromagnet between
the diaphragm and carbon resistance; several types in which the diaphragm short
circuited the coils of a wire wound resistor; and an "inertial" telephone. Edison was
awarded patents on some of these but did not apply for a patent on the capacitance

---

[51]TAEB 3:765.

[52]Jehl, *Menlo Park Reminiscences*, 1, 143–54.

(or condenser) microphone, a design that varied the voltage in the circuit by varying the electrical charge between insulated plates a microphone operating principle that has replaced the carbon microphone in most telephones including mobile phones.

Examination of these patents today suggests that most of the devices had little likelihood of commercial viability. It is likely therefore, that Edison was using the patents strategically. By obtaining patents on a wide range of alternative approaches, he could prevent or at least inhibit other inventors from developing them into competitors for the device he favoured.

2.9   January 1877: Carbon Enters the Telephone

Initially Edison worked with transmitters using either Reis style switches or with variations on the concept of liquid variable resistance in his undated sketch (Fig. 2.6), but in October 1876 he took a different path, testing one arrangement that used metallised felt to short circuit a thin plumbago (mineral graphite) coating on hard rubber and another in which a plumbago coated Arkansas oilstone dipped into water (Fig. 2.12).

These experiments were Edison's first use of carbon in a transmitter. He did not return to the telephone until January 1877 when he sketched a transmitter that, for the first time, worked by varying the pressure on carbon rather than varying the amount of carbon included in the circuit. In this design, the variable resistance was created by the ends of platinum points attached to the diaphragm vibrating in loose carbon in a dish, a variation on his partially successful electrolyte transmitter of July 1876 (Fig. 2.13).

The performance of this design was a mixture of failure and success but Edison said that it prompted him to recall the vibration sensitivity of the carbon rheostat he had devised in 1873 following his failed Greenwich demonstration. Although the drawing of the carbon rheostat (Fig. 2.3) does not show means for varying the pressure on the carbon, in his testimony Edison's associate Charles Batchelor,[53] said of the device that "the resistance was greater or less according to the compactness of the carbon".[54]

---

[53]Charles Batchelor (1845–1910) was English born textile mechanic who became Edison's primary associate in invention for 20 years. Batchelor joined Edison's Newark American Telegraph Works as a machinist in 1871 and by the summer of 1873 was assisting Edison in inventing. Later Batchelor held significant positions in Edison enterprises, living in Europe for 3 years building Edison's electric lighting systems. Batchelor's careful and meticulous approach to inventing complimented Edison's fountain of ideas and together they made an effective team. Edison recognised Batchelor's contribution by allocating him a significant proportion of the royalties from patents they worked on together. Batchelor left Edison after Edison's disastrous magnetic ore extraction venture in which Batchelor lost his substantial investment.

[54]TAED TI1:91.

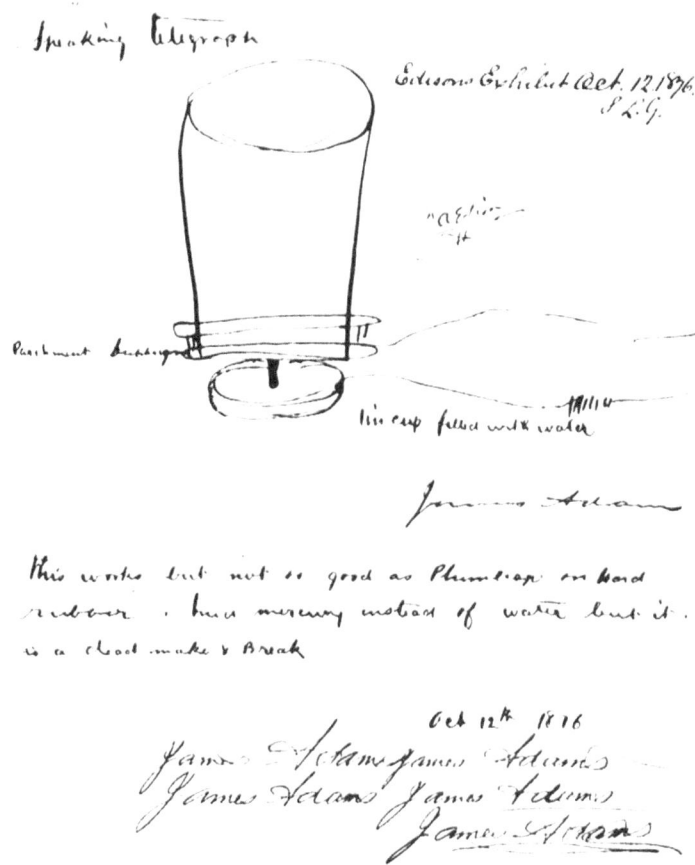

Fig. 2.12 Edison's telephone sketch of 12 October 1876 using electrolyte as variable resistance. (TAED TI2:43)

Jehl describes a different kind of rheostat that seems to fit Batchelor's description better. The device Jehl described consisted of "fifty or more silken discs filled with fine particles of graphite [between them]".[55] A micrometer screw was used to vary the pressure on the disks, producing resistances of from 400 to 6000 ohms. Whatever the design, the rheostat proved a failure for its intended purpose because the slightest bump or vibration caused its resistance to change, knowledge that Edison recalled and exploited in January 1877. While at this stage Edison did not have a theory for the pressure and vibration sensitivity of the carbon rheostat, he knew empirically how it behaved.

Following this recollection, Edison and Batchelor started a period of intensive work on the telephone. In February 1877, they tested a variety of designs in addition to those with a carbon variable resistor. These included one using capacitance

[55]Jehl, *Menlo Park Reminiscences*, 1, 113.

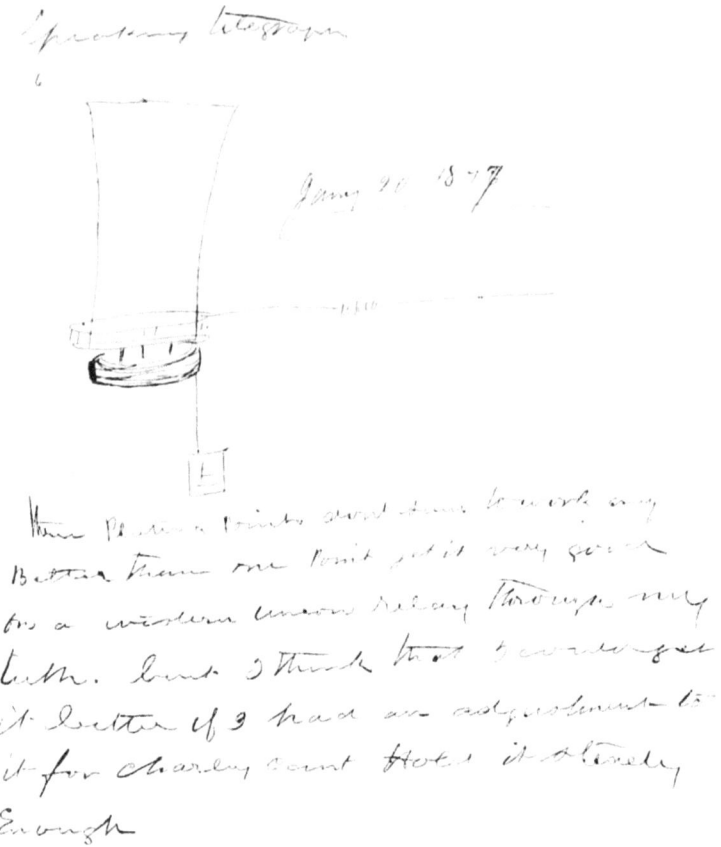

Fig. 2.13 Edison's telephone sketch 20 January 1877 using granulated carbon. (TAED TI2:45)

instead of resistance, another using moist paper and another using multiple resistors and diaphragms. Although these new operating principles failed, by this stage the carbon based transmitter had been developed to the point where Edison could declare that "I used it on a telegraph wire and transmitted and reproduced articulate speech with such a degree of success that it was capable for business purposes".[56] In mid-March, Edison demonstrated this carbon microphone to Western Union using a line between Menlo Park and New York City, a distance of about 40 km.[57]

On 1 April 1877, Edison sketched a design that used pressure applied to blocks of plumbago to vary the resistance (Fig. 2.14). Edison was now working with the theory, based on observation of the 1873 carbon rheostat that the resistance of carbon

---

[56]TAED TI1:83.
[57]TAEB 3:874n1.

Fig. 2.14 Transmitter using
blocks of plumbago.
Edison's notes are,
"plumbago or other inferior
conductor" "Speak" "stiff"
"Line" and "April 1 1877".
The signatures are those of
Edison, Adams and
Batchelor. (TAED TI2:73)

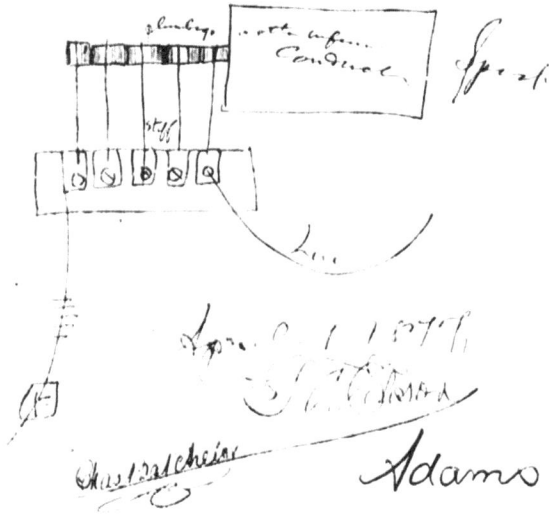

varied with pressure. Although he had a workable theory, he had not yet developed an effective way of exploiting it. Despite this, he was making progress and noted on 4 April that "we can get everything right except the lisps & hissing part of speech such as "sh" in shall = get only .o. in coach".[58]

In addition to developing a telephone transmitter, Edison was also developing a clearer picture of how it should perform. He had previously identified the need to reproduce what he called hissing sounds (sibilants) such as "ch" and "s", but this is the first instance in which he identified performance on vowels as a problem. (His patent application of 28 August, 1877 was specifically intended to deal with the problem of reproducing what Edison called "hissing consonants".[59]) On 27 April 1877, he applied for his first telephone patent based on the principle of varying the pressure applied to carbon. It was this patent that was contested in the courts, and resolved in Edison's favour in 1892.[60]

In the same 1 April notebook entry Edison also lists "inferior conducting materials" could be used in variable resistors including black manganese oxide, anthracite coal, bituminous coal and plumbago. On 10 April, he sketched a relay, which he later patented, that exploited the variable resistance of these materials to amplify current.[61] Relays of this type were used to increase the distance over which telegraph signals could be transmitted.

---

[58]TAEB 3:882.

[59]Thomas A Edison. Speaking-Telegraphs [3]. US Patent 203,015, filed 28 August, 1877, and issued 30 April, 1878.

[60]Speaking-Telegraph [2].

[61]TAEB 3:885.

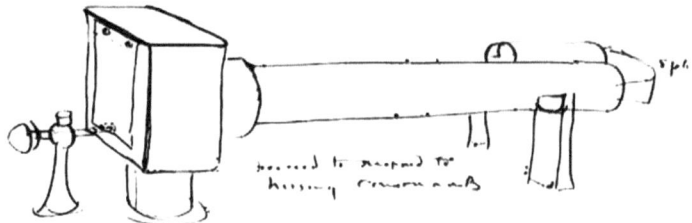

Fig. 2.15 Telephone transmitter (microphone) design using tubes of different or adjustable length. Note is "free reed to respond to hissing consonants". (TAED NV11:61)

On 26 May 1877, still struggling to transmit sibilants, Edison proposed a hypothesis about a characteristic of speech. This was that sibilants had "an exceedingly low rate of vibrations pbly 10 per second & weak at that".[62] From this, he proposed several solutions including telescopic speaking tubes, twin tubes of different lengths and a design that used a free reed (Fig. 2.15).

On 25 May, Edison sketched another arrangement using the same principle as Fig. 2.14 but with plumbago blocks pressed onto both sides of the diaphragm. It was this arrangement, rather than the one in Fig. 2.14, that is illustrated in the patent that Edison eventually received, despite his legal argument for priority being based on the arrangement in Fig. 2.14.[63] This is also one of the first drawings showing the horn shaped mouthpiece used in his final arrangement rather than the simple cylindrical speaking tube adapted from his acoustic telegraph apparatus. At this stage, Edison was dealing with two different problems arising from the failures he had identified: finding a suitable variable resistor material and finding the best physical arrangement to exploit the material.

The search for a suitable variable resistor material went on intermittently for months with hundreds of substances, mainly mixtures, being tested. In June 1877, Batchelor and Adams worked through several hundred mixtures of plumbago and other substances in search of a better variable resistor material, recording their relative success and failure of each at reproducing whispering, whistling and other sounds. On 18 June they noted in relation to one mixture, numbered 151 (plumbago mixed with ground rubber) "This is the one Edison likes!!!".[64]

---

[62]TAEB 3:921.

[63]Edison. Speaking-Telegraph [2].

[64]TAEB 3:941.

2.10   July 1877: Fluff

In June and July 1877, development progressed with variations in configuration including rotating wheels, a weighted diaphragm, and attachments intended to capture sibilants. On 17 July 1877, the laboratory notebook included the jubilant comment "Glorious = Telephone perfected this morning at 5 am = articulation perfect got 1/4 column newspaper every word".[65] It is a typical optimistic Edisonian declaration because, despite the confident tone, the claimed perfection proved premature and work on the telephone transmitter continued. On 30 July, Batchelor began testing a new material, a mixture of silk fibres and plumbago, they named "Fluff", Fluff becoming the preferred resistance material over the following months. On 20 August, Edison and Batchelor demonstrated a Fluff based transmitter to Western Union using a line between two buildings in New York.[66]

Fluff was successful as a resistance material but failed in a new and unexpected way. Edison and Batchelor noticed that the resistance of Fluff increased over time, discovering the cause was the carbon separating from the silk fibres, a problems that caused them to expend considerable effort to solve. Searching for a solution, on 10 August, Edison explored the limits of the Fluff mixtures, noting that "the conducting fibre can be dispensed with in my telephone & its equivalent substituted ie the clean fibre may have a semiconducting substance included in its folds & that will work Even loose plumbago or equivalent will work this".[67] Despite this discovery, Edison did not pursue pure plumbago but persisted with the Fluff mixture. Ever optimistic, he wrote on 24 September, "We have got the Fluff biz dead to rights".[68] As with his earlier declaration of perfection, this claim proved premature and Fluff separation persisted.

The same day, still testing mixtures, Edison observed that one rapidly went out of adjustment. He did not immediately abandon the mixture but sought an explanation for the failure, proposing that it was due to the electric current heating the material.[69] Tested and confirming this hypothesis, he subsequently exploited the phenomenon (which he referred to as the Tasimeter principle) in a number of devices. These included the Micro Tasimeter, an instrument he devised to detect very small changes in temperature and another that regulated current to in one of his early incandescent lamps.[70] Edison used the Micro Tasimeter to observe a solar eclipse in Wyoming in July 1878.

On 22 October 1877 Edison returned to testing mixture 151 from June (plumbago mixed with ground rubber) and, on 26 October, lamp black mixed with rubber. Even at this stage, with so much effort directed to a carbon transmitter, Edison was still

---

[65]TAEB 3:968.
[66]TAEB 3:1016n1.
[67]TAEB 3:1005.
[68]TAEB 3:1067.
[69]TAEB 3:1068.
[70]TAEB 4:1289n2.

exploring alternatives, including galena and bismuth sulfide as semiconducting materials and radically different operating principles including a telephone that generated its own current as Bell's original did.[71] Around this time Edison appears to have devised a new theory for the way in which carbon responded to sound, because Batchelor recorded on 9 November that Edison had "found out that plumbago does not alter its resistance by pressure as we at first thought, but the increased pressure made better contact".[72] There is no record of how Edison came to this new theory but crucially, the new theory implied that the contact area between the particles was critical, so the more finely divided the material, the more effective it would be. With this knowledge Edison began using lamp black (extremely fine soot from oil lamps) as the new resistance material. Writing to a business associate on 22 December, 1877 Edison mentioned that he was now using lamp black alone (without silk fibres) and had found that by purifying it he could make the transmitter much louder.[73] This new theory (that resistance depended on contact area and not pressure) was Edison's last major step in his search for a suitable resistance material.

On 4 February 1878, Edison successfully demonstrated his lamp black carbon microphone between Menlo Park and Philadelphia, a distance of 210 km, with one witness reporting "I recognised your voice instantly".[74] A later test on 12 March between Philadelphia and New York proved a partial failure because a critical component, a rubber tube between the diaphragm and carbon, had lost its flexibility.[75] Edison had introduced the rubber tube (Fig. 2.16) in the belief that a flexible material was necessary to transmit the vibrations, an error that seems to have been a consequence of using a drum as his theoretical model. To overcome the failure of the rubber tube, he replaced it with a metal spring but found this "gave a musical tone" (probably caused by the spring resonating at an audible frequency). Edison then tried successively stiffer springs to reduce the resonance, eventually substituting a rigid brass tube. Finding this gave the best results of all, he concluded "the whole thing was one of pressure only, and that it was not necessary for the diaphragm to vibrate at all". (Edison was not strictly correct in this. The diaphragm did vibrate but in the frequency range he was working with the amount of vibration would have been imperceptible, unlike that in the head of a drum.)

To test his hypothesis, he tried a stiff diaphragm (1.6 mm thick) and joined the diaphragm and carbon rigidly together: "Upon testing it I found my surmises verified; the articulation was perfect and the volume of sound so great that a conversation carried on in a whisper three feet from the telephone was clearly heard and understood".[76] This rigid coupling between the diaphragm and carbon variable resistance became at first a brass tube then in the final form, an aluminium

---

[71]TAEB 3:1095.

[72]TAEB 3:1107.

[73]TAEB 3:1154.

[74]TAEB 4:1194.

[75]TAEB 4:1251.

[76]TAEB 4:1251n3.

Fig. 2.16 An early Edison transmitter design showing a rubber tube (f) between the diaphragm (g) and resistance (D). (Thomas A Edison. Speaking-Telegraphs [1]. US Patent 203,013, filed 13 December, 1877, and issued 30 April, 1878)

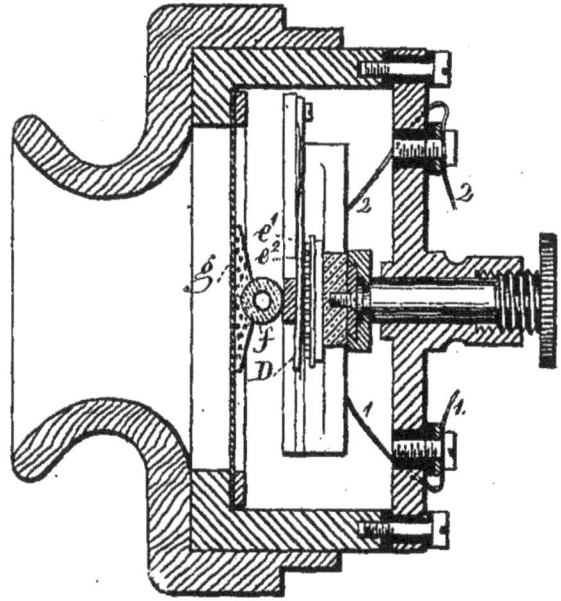

button (labelled "A" in Fig. 2.1). With this successful design, Edison began commercial negotiations with both Western Union and Bell Telephone. Western Union was successful and on 31 May 1878, Edison assigned his telephone patents to the company in exchange for royalties of $6000 a year for the duration of the patent. With this in hand, Western Union immediately began manufacturing Edison's carbon microphone for its telephone system.[77]

2.11 Failure as a Tool

This sketch of the major steps in inventing the carbon microphone illustrates Edison's use of failure as a tool to produce a successful invention. The quotations that began this chapter show Edison's positive attitude to failure but also that he believed the value of failure was primarily to identify "things that won't work". These quotations come from the latter part of his life when Edison was past his inventive peak and are perhaps more intended to enhance his public image as the yokel inventor who made good, than to provide an accurate account of his work.

---

[77]Paul B Israel, *Edison: A Life of Invention* (New York: John Wiley, 1998), 141.

The reality of Edison's use of failure is far more complex. Had Edison only used failure as a source of negative examples, he would have taken his failed Greenwich demonstration as a lesson that the "thing couldn't be done that way" (using a coiled cable) and sought other ways of demonstrating his automatic telegraph. While he did this, he did much more with the failure. The experience spurred him into seeking to understand more about the phenomenon that caused it, induction. The 6 months spent researching induction gave Edison a deep understanding of induction and its effects, knowledge he later applied in many situations and inventions, including a circuit he patented to overcome induction in long parallel telephone cables and a device he sold as "Edison's Inductorium" that delivered electric shocks as a medical treatment or for entertainment.[78,79]

Edison's anecdotes illustrate two common beliefs about failure. The first is expressed by Edison's associate who saw failure as a disappointment and a setback. This understandable response reflects the commonly held view that only success is important. For people who hold this view, success is of value because it can be built upon success to create more successes while failures are just a waste of time and effort. Had this been true for Edison, he would have progressively refined the telephone transmitter by seeking successes and the knowledge that came from them. Some have taken this further, arguing that the progress of technology parallels biological evolution with more successful technological artefacts displacing less successful ones.[80] (As used here, an artefact is "an object which has been intentionally made or produced for a certain purpose".[81])

The second belief is expressed in Edison's "crowning experiment" anecdote is that failure, even thousands of failures, can be viewed positively because from them we learn what cannot be done. Despite its positive tone, such a view is little more than a consolation. It differs from Edison's associate's view only in degree because it values failure only as a source of negative examples. Edison offers nothing else to be gained from failure so it seems that he, like his associate, believed that most knowledge of value comes from success. While these two positions are understandable, common sense responses, their emphasis on success limits the value failure, value that goes well beyond identifying that "the thing couldn't be done that way". Edison the inventor is better understood through his responses to failures than through his response to his successes. To demonstrate this I will summarise the ways in which Edison used failure as a tool in inventing.

---

[78]Circuits for Acoustic or Telephonic Telegraphs. US Patent 203,019, filed 21 February, 1878, and issued 30 April, 1878.

[79]TAEB 2:434, 2:435.

[80]for example George Basalla, *The Evolution of Technology* (Cambridge, Cambridgeshire: Cambridge University Press, 1988).

[81]Risto Hilpinen, "On Artifacts and Works of Art," *Theoria* 58 (1992); "Belief Systems as Artifacts," *The Monist* 78, no. 2 (1995). In common scientific usage, an artefact can also be an observed defect or aberration. Both this, and the meaning used here, share a common origin in referring to something that does not occur naturally.

2.11.1   Failure as a Source of Negative Examples

The first and most obvious way Edison used failures was to accumulate a set of negative examples, "everything that don't do it" as Edison described it. In doing this, he adopted an approach similar to that advocated by computer scientist Janet Kolodner who proposes solving future problems by assembling and interrogating a corpus of past failure and success cases. She notes that,

> A reasoner whose cases cover more of the domain will be a better reasoner than one whose cases cover less of the domain. One whose cases cover instances of failure as well as success will be better than one whose cases cover only success.[82]

While the use of failure in this way has value, it is severely limited since failure cases provide only negative knowledge, things "that don't do it". Seen this way, failures provide no knowledge of what to do to achieve success. For such positive knowledge, Kolodner relies on past successes but in Edison's development of the telephone transmitter, there were few of these. Had Edison confined himself to past successes it is doubtful that he would have produced much of value.

2.11.2   Using Failure Through Trial and Error

Apart from using failure to find "everything that don't do it", Edison is perhaps most famous for his use of failure in the form of trial and error. He used it so extensively that trial and error is sometimes referred to as the Edisonian Method. Chapter 10 examines trial and error in detail, but for the present, it is important to note that Edison's use of trial and error was not a random, aimless process conducted in the hope that something would eventually appear. Instead it was one in which each trial failure, success or mixture of failure and success, gave a direction to the next trial, enabling the process to converge towards a successful solution. For Edison, trial and error was a process in which the analysis of failure was central. Patterns that emerged from trial and error could be developed into systems of regularities, repeatable patterns that Edison could use to predict future behaviour of phenomena in a manner analogous to scientific theories when no relevant scientific theories existed.

2.11.3   Failure as a Source of New Phenomena

A more enterprising response to failure can be seen when Edison was confronted with unexpected results, a number of instances of which occurred as he developed

---

[82]Janet L. Kolodner, *Case-Based Reasoning* (San Mateo, CA: Morgan Kaufmann Publishers, 1993), 7.

the carbon microphone. Edison's response to these was to explore the anomalies in search of new phenomena since Edison viewed new phenomena as potential sources of new inventions. Edison recorded these anomalous phenomena as he observed them, sometimes pursuing them in search of new inventions, sometimes just recording them as something that seemed interesting. On 14 January 1877, while working with acid solutions he made a notebook entry headed "Phenomenon" describing the effects of acids on paper.[83] This phenomenon went no further than the observation of something curious but on 10 August 1877 he turned the apparent failure of a test microphone into the discovery of the Tasimeter principle, exploiting it later in a number of inventions. Similarly, the failed Greenwich demonstration led to Edison's exploration of induction and to inventions including the Inductorium.

It is possible to trace a chain of these exploited phenomena, linking unexpected results observed while developing one invention to the development of another: the failed Greenwich demonstration of the recording telegraph led to invention of the Inductorium and carbon rheostat; observation of the failure of the carbon rheostat fed into the telephone, which led to the Tasimeter and carbon relay.

Hughes notes "[Edison] sought the stimulating effect of the interaction between the system components he was developing. Imbalances identified the requirement of additional invention".[84] Hughes's "imbalances" can be interpreted as anomalous results that arise when components do not work together as anticipated.

2.11.4   Failures Provide Direction

Hughes uses a military metaphor, the reverse salient, to describe the approach to inventing used by independent inventor-entrepreneurs.[85] In a reverse salient, the presence of the enemy in the salient holds back an advancing of army (Fig. 2.17). In the inventing situation, an unsolved problem in one particular area can hold back the development of a whole branch of inventions. Hughes notes that the inventor who identifies and solves such reverse salient problems can advance a technology more than one who works with the more advanced parts of the technology.

According to Hughes "Once independents [independent inventor-entrepreneurs] embarked on the voyage of inventing a system, there were beacons all along the way. They concentrated on the sequence of reverse salients as the appropriate problem choice".[86] According to Hughes, these reverse salient "beacons" directed the inventor's effort towards the aspects of the invention holding back the solution. We can

---

[83]TAED NV08:111.

[84]Thomas P Hughes, "Edison's Method," in *Technology at the Turning Point*, ed. William B Pickett (San Francisco: San Francisco Press Inc., 1977), 9.

[85]*American Genesis: A Century of Invention and Technological Enthusiasm 1870–1970*, 2nd ed. (Chicago: The University of Chicago Press, 2004), 53–95.

[86]Ibid., 73.

Fig. 2.17 A military
reverse salient

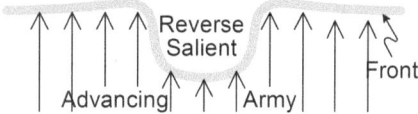

understand Hughes's beacons as crucial failures that impede overall development, so the reverse salient process moves from failure to failure solving each on the path to overall success. For Hughes, the independents used reverse salients both to select projects and to direct their solution process. Hughes does not develop his reverse salient concept in relation to the inventors' methods beyond this, but gives considerable attention to it in the context of his approach to large technological systems in which he identifies reverse salients as pointing the independents towards new invention projects.

2.11.5 Hypotheses and Failure

At many points during Edison's development of the carbon microphone, we saw him seeking to understand what has gone wrong, asking "why?" questions, answering them either explicitly in his notes or, more often, by implication in his sketches of new devices. Although Edison admitted to using empirical methods like trial and error to solve chemical problems, he observed, "when it comes to problems of a mechanical nature, I want to tell you that all I've ever tackled and solved have been done by hard, logical thinking".[87] This logical, analytical process led to hypotheses followed by experiment, the hypotheses becoming new solutions to test in new experimental models and, if verified, theories. Examples in the development of the carbon microphone include his hypothesis that different parts of the diaphragm produced different tones; that a resistance mixture degraded due to heating; and his assertion that "Articulate speech consists of two parts Musical sounds and clang sounds, the first varying in pitch and volume with enormous rapidity, while the latter is composed of vibrations having no definite relation to one another".[88]

These hypotheses were tested experimentally, for example Edison's hypothesis about different parts of the diaphragm responded to different tones. If the hypothesis was not validated it was discarded, but significantly, it became more than just a negative example. Discarded or not, each of these hypothesis-test sequences helped Edison build better theories of how the device and its components functioned. Occasionally such a hypothesis led to a critical change in a theory, marking a turning point in the development process, as when Edison abandoned the theory that the resistance of carbon varied with pressure for a new theory that resistance varied with contact surface area. This new theory implied that he needed a substance with a large

---

[87]Dyer and Martin, *Edison, His Life and Inventions*. 609.
[88]TAEB 4:1315.

surface area so he turned to carbon black. Although Edison stated his diaphragm hypothesis explicitly, in many situations he relied on an observed empirical relationship. In this way, he was able to transfer the observed vibration sensitivity of his 1873 carbon rheostat to the telephone without having a theory for the vibration sensitivity of carbon.

2.11.6 Failure Points to Ways of Changing Devices

A common way in which Edison used knowledge from failures was to identify ways to change the device for the next attempt. When he identified multiple ways to change it, the result was a branching process. The greater the branching the more likely they would reveal a new phenomenon that he could pursue to produce another invention through more failures and successes. Each of these branches has potential to branch in the same way, producing a network of inventions linked by the exploitation of unexpected results.

Anomalous results are the exception however. A far more common result of failure is that device does not work at all or works poorly. In January 1877, Edison was trying to produce a variable resistance by short circuiting plumbago coated on rubber. This worked poorly but introduced carbon in the form of plumbago into his invention process. Edison then combined carbon with the partially successful electrolyte transmitter from the previous October (Fig. 2.12) by replacing the electrolyte with loose plumbago powder (Fig. 2.13). Variations on the principle of short-circuiting a resistance subsequently became the basis for a series of patent applications. In these instances, failure became the motivation to seek other solutions.

For Edison, failure demanded an explanation and the explanation led him to propose new solutions. Even if the solution was not completely successful (which was usually the case), Edison used it to generate more hypotheses and potential solutions to build and test. His notebooks indicate that most solutions failed, but occasionally a partial success led to a more fruitful line of development. Failure causes branching of the artefact creation process and hence potentially increases both the number of artefacts produced and the variations in possible solutions. In most instances when Edison changed something, it was because a device failed in some way. "Change for change's sake" is rare for Edison.

This means, firstly, that the process of creating artefacts is not a linear one but one with much branching. Secondly, it implies that this branching is the product of a sequence of failures, not of successes. Thirdly, and a consequence of the first two, failure and overcoming it are so valuable that they can be seen as primary drivers of change in artefact creation. That is to say, Edison worked by seeking the ways in which his devices failed. Success had its place (and obviously success was Edison's ultimate objective), but it was failure that gave his process its direction.

2.11.7   Failures Suggest New Possibilities to Build and Test

One aspect that distinguished Edison from those who worked with him was his ability to generate a large number of ideas and to build models of these for testing. Having identified the limited geographic range of Bell's telephone as a critical failure, Edison generated several ideas that he sketched for later building and testing (Fig. 2.6). Although Edison's sketches appear simple, even crude, the physical models they became were not.

A striking feature of his laboratory notebooks is the quantity of sketches that seem to have gone through this process. Dyer and Martin quote one of his associates as saying "Edison can think of more ways of doing a thing than any man I ever saw or heard of."[89,90] In November 1877, Edison produced 116 dated laboratory notebook pages containing about 400 sketches. Based on the notebook sketches and surviving models, I estimate that building and testing each would have taken several hours to several days, depending on whether the model was cannibalised from existing models (Edison's favoured method) or built from scratch. Not all of Edison's sketches were built and tested but even if the number was a small percentage of his total output (400 in this one month), it represents a substantial commitment in time, labour and materials. This, no doubt, contributed to Edison's reputation for driving himself and his associates hard. He needed to in order to work through so many ideas.

2.11.8   Failure Provides Motivation

In Edison's case, one of the most significant uses of failure was as motivation. Edison had no need to modify his successes but he certainly needed to fix his failures. It also appears that for Edison (and apparently his associates) failures were not so much trials to endure as intriguing puzzles to solve. Edison tended to become obsessed with the pursuit of new problems, notoriously ignoring sleep, financial cost and, to Tesla's disgust, personal hygiene. It is not hard to picture this group of young men, closeted in rural Menlo Park, having fun, creating new and amazing inventions while being paid for it. As one of Edison's associates, Charles Clarke described it "Here breathed a little community of kindred spirits, all in young

---

[89]Dyer and Martin, *Edison, His Life and Inventions*. 603.

[90]Frank Dyer (1870–1941) New York lawyer, Frank Dyer was for many years, general counsel to Edison, the Edison Laboratory and Edison companies and first president of Thomas A. Edison, Incorporated. With electrical engineer, Thomas Commerford Martin, Dyer wrote a widely read authorised biography, *Edison, His Life and Inventions* (1910).

manhood, enthusiastic about their work, expectant of great results, often loudly explosive in word, emphatic in joke, and vigorous in action".[91]

2.12   Calculation and Experiment

Some have dismissed Edison's approach of building and testing many prototypes. Nicola Tesla described it as "inefficient in the extreme" because "just a little theory and calculation would have saved him 90 per cent of the labour".[92] Although Tesla worked for Edison's organisation, he never worked directly with him, so Tesla did not have firsthand knowledge of his methods. However, there is some truth in his observations on Edison's lack of mathematics, Edison's notebook entries for the carbon microphone containing very little in the way of calculation. The November 1877 notebook pages discussed previously may contain hundreds of sketches, but include only a handful of calculations and even these use only basic arithmetic.

The evidence of Edison's notebooks and the reports of those who worked with him suggest that he built and tested physical models almost exclusively whereas other inventors used a mixture of mathematical and physical models. Israel argues that Edison's preference for working with models was in keeping with nineteenth-century Anglo-American scientific and technical practice of which Michael Faraday and William Thomson (Lord Kelvin) are two exemplars.[93,94] Faraday, perhaps the nineteenth century's greatest experimentalist, used very little and very simple mathematics. Thomson also created a significant body of theoretical knowledge. Like Faraday, he preferred to work experimentally. Unlike Faraday, he did this despite being a gifted mathematician. Both Faraday and Thomson believed that direct experimental experience produced knowledge not available from a theoretical-only approach.

Despite his preference for experiment, Edison was economical in his approach, observing that "A good many inventors try to develop things life-size, and thus spend all their money, instead of first experimenting more freely on a small scale"

---

[91] quoted in Bernard S Finn, "Working at Menlo Park," in *Working at Inventing: Thomas a Edison and the Menlo Park Experience*, ed. William S Pretzer (Dearborn, Michigan: Henry Ford Museum & Greenfield Village, 1989), 44.

[92] New York Times. "Tesla Says Edison Was an Empiricist." *New York Times*, 19 October 1931, 25. ProQuest Historical Newspapers The New York Times (1851–2003).

[93] Israel, *Edison: A Life of Invention*, 95–96.

[94] William Thomson (Lord Kelvin) (1824–1907) Scottish mathematician, scientist and inventor. In science, Thomson was noted for his work in thermodynamics, including formulation of an expression of the Second law of thermodynamics, the Joule-Thomson effect and the concept of absolute zero temperature. He was also successful as an inventor, building a considerable fortune through his contributions to telegraphy including the stranded cable and mirror galvanometer, instruments, improvement to marine compasses and, with his brother the engineer James Thomson, a harmonic analyser used in meteorology.

and that "When there's no experimenting there's no progress. Stop experimenting and you go backward. If anything goes wrong, experiment until you get to the very bottom of the trouble".[95]

Like Edison's associate quoted earlier, Tesla may have thought that Edison's model building approach and its many failures were inefficient but Edison did not. His use of failure as a tool meant that the failure of an experiment or test did not mean a failure to build relevant knowledge.

In addition to the knowledge gained from failures, by working on models of the whole artefact Edison was able to address all parts together and so deal with the messy uncertainties that their interactions produced. The analytical approach advocated by Tesla requires decomposing problems into independent variables and components. It may make analysis and quantification easier, but gives less information about the behaviour of the whole device as a complex system. Edison's approach may appear inefficient but in part it was time consuming because it highlighted failures in aspects of the model that the analytical approach could not. If the objective was to create successful inventions, then in Edison's hands and with the support of skilled associates like Charles Batchelor, it proved an effective strategy. Another crucial associate was John Kruesi, a Swiss born clock and instrument maker who joined Edison in Newark in 1872.[96] Kruesi translated Edison's rough sketches into working models for many of Edison's inventions at Newark and Menlo Park.

Edison progressively employed more highly qualified staff to undertake complex mathematical analyses. These included Francis Upton, who had studied under Helmholtz, and Frank Sprague, trained in engineering at the United States Naval Academy, Annapolis. Jehl may have lauded Edison's inventive genius in his 1937 book but in 1913, he complained privately that Edison could "accumulate such wealth with such little real knowledge, a man that cannot solve a simple equation".[97] The analytical and mathematical skills of Jehl and Tesla may have improved aspects of Edison's inventions but they did not increase his output of patent applications. While Tesla may have thought that Edison was "inefficient in the extreme", as an inventor, Edison's patent output peaked between 1880 and 1883, before Tesla arrived and when he had few mathematically trained associates. Mathematics may have improved inventions but in Edison's case, it did not increase his output of inventions.

---

[95] quoted in Dyer and Martin, *Edison, His Life and Inventions.* 617.

[96] John Kruesi (1843–1899) Swiss American instrument maker and Edison associate. Kruesi was born in Switzerland being apprenticed to a clock and instrument maker. After emigrating to America, he joined Edison in Newark in 1872, moving with him to Menlo Park where he headed the machine shop that occupied much of the ground floor of the laboratory. Kruesi made many of the models for Edison's inventions at Newark and Menlo Park, and oversaw the installation of Edison's first electric lighting station in New York. Kruesi became general manager of the Edison Machine Works.

[97] quoted in Hughes, *American Genesis*, 91.

Despite Tesla's claim that Edison used no theory or calculation, Edison's papers reveal that he did employ economic and scientific analyses when appropriate. For example, a key part of Edison's development of an economically viable electric lighting system was his use of Ohm's and Joule's laws combined with economic analysis to conclude that the optimum resistance for the filament of his incandescent lamp should be around 100 ohms.

In the search for a suitable microphone resistance material, Edison had no equivalent theory available to solve his problem because at the time there was no theory to predict the behaviour of semiconducting materials under these circumstances. Consequently, Edison and his associates embarked on the long process of testing candidate materials using a device that Kruesi built to measure the pressure resistance relationship of each material.[98] While Edison's use of trial and error is often regarded negatively, as discussed in Chap. 10, trial and error is of considerable value in situations where there is no relevant theory, as it may be the only way to arrive at a solution, regardless of the training or experience of the practitioner making use of it.

2.13　Seeking Failure

Edison's work on the carbon microphone shows that he did not merely encounter failures; he actively pursued them, pushing his devices to discover new ways in which they failed. Not only did he try alternative materials and details, but in testing the telephone transmitter, he sought more and more demanding conditions, over greater and greater distances, beginning in his Menlo Park laboratory, then between Menlo Park and New York, eventually testing it over several hundred kilometres between New York and Philadelphia. Had his objective been to achieve any degree of success, he may have stopped after one of his early successful transmissions.

It is clear from the development of the carbon microphone that the image of Edison starting work on an invention with a brilliant "eureka moment" then developing it through trial and error tinkering is inadequate to the point of being a parody. Edison famously said, "Genius is about two per cent inspiration and ninety-eight per cent perspiration".[99] The evidence suggests that in Edison's case at least, 98% perspiration might be an underestimate. It is more apt to see his way of working as a methodical grind with periodic (and thoroughly tested) minor breakthroughs. In retrospect, we can see that Edison did have moments of inspiration, but his notebooks show that in most cases he approached a new idea tentatively and only took it up in earnest when he had thoroughly tested it and eliminated alternatives. Paradoxically, on the occasions when Edison declared that he had made he had a

---

[98] Jehl, *Menlo Park Reminiscences*, 1, 117.

[99] quoted in Francis William Rolt-Wheeler, *Thomas Alva Edison*, True Stories of Great Americans (New York: Macmillan, 1925), 92.

breakthrough, as when he wrote that he had "got the Fluff biz dead to rights", hindsight often showed him to have been mistaken. Edison's case suggests that it is hindsight that enables us to identify eureka moments, not the euphoria of the moment. The role of failure in the invention process means that we cannot be sure we have a eureka solution until we have tested it, found it to be successful and identified its limitations.

The many similarities between Edison's approach and the processes used by contemporary technologists suggest that failure also plays a central role in the creation of contemporary technological artefacts. Perhaps most importantly, this parallel implies that, like Edison, contemporary technologist get direction from failure and that, paradoxically, the success of an artefact reflects its creators' success in identifying and overcoming failures.

Chapter 3
Failure and Success

3.1 Edison Takes a Break

December 1877 to May 1878 was a hectic period for Edison, culminating in the sale of his carbon microphone patent rights to Western Union. The period began spectacularly in December 1877, with Edison's demonstration of a revolutionary invention, the phonograph. The Phonograph made Edison a celebrity but led to a busy schedule touring the country demonstrating it while still working long hours at inventing. By May, Edison was haggard and ready for a break, which fortuitously came in the form of an invitation to join a scientific expedition to Rawlins, Wyoming to observe a solar eclipse in July 1878.

Edison's role in the expedition was to use his Micro Tasimeter to measure the temperature of the Sun's corona. The long journey also gave Edison the opportunity to discuss science and his inventions with his fellow solar observers and an opportunity to be a tourist. As with so many things that Edison did, his approach to tourism was unconventional. Through his connection to Union Pacific Railroad president, Jay Gould, Edison received a letter instructing Union Pacific officials to give him whatever he wanted. What Edison wanted was a cushion and permission to travel on locomotive cowcatchers. The result was that he travelled most of the way from Omaha to the Sacramento Valley on this novel viewpoint.[1]

In the end, last minute problems with the Micro Tasimeter meant that his scientific contribution to the expedition was marginal but the break from his heavy workload gave him the opportunity to consolidate some of his ideas and develop new ones. Of these, the most significant was electric lighting, Edison throwing himself into the problem the day after his return home.

Just as Edison took a break to consolidate his ideas and develop new ones, this is an appropriate point to pause and develop the theoretical approach to failure that will be applied in later chapters. This chapter introduces concepts related to failure

---

[1]Israel, *Edison: A Life of Invention*, 162.

© Springer Nature Switzerland AG 2019

I. Wills, *Thomas Edison: Success and Innovation through Failure*, Studies in History and Philosophy of Science 52, https://doi.org/10.1007/978-3-030-29940-8_3

including success criteria, success clues and success frameworks. Later, Chap. 6 will look at catastrophic failures and their relationship to the kinds of failure we saw in Edison's development of the carbon microphone.

3.2   Failure

The word failure has been used many times to this point but before looking at its implications we need to clarify what it means in the present discussion.

When we say that something has failed, we are saying that it has not meet one or more conditions or criteria for success (*success criteria*). As simple as this statement is, it has several important implications. The first is that there is no limitation on who sets success criteria, what count as success criteria or when a determination of success or failure is made. Something that was successful in the past may not continue to be successful. Edison's carbon microphone was the standard in the telephone industry for a century but the introduction of electronics, particularly amplifiers, into telephones around 1980 meant that its primary characteristic, variable resistance was no longer needed and other criteria appeared including an emphasis on small size. The result of the application of these new success criteria is that this once extremely successful device is now essentially obsolete.

The second implication of this definition is that does not limit the form success criteria can take. Often success criteria are explicit but they can also be implied. The definition used here is broader than some others. The US Department of Defense defines failure as "The event, or inoperable state, in which any item or part of an item does not, or would not, perform as previously specified".[2] By requiring criteria to be "previously specified", this definition both excludes implied criteria and requires all success criteria to be stated in advance. This may suit the needs of the Department of Defense but it does not reflect failures encountered in many situations. The development of the carbon microphone revealed many instances of implicit success criteria being recognised retrospectively, success criteria that rendered Edison's developmental carbon microphone designs failures. Among the implied criteria for Edison's microphone was that it have a usable operating life. When Fluff-based instruments deteriorated after only few weeks, they were judged failures against this implied success criterion.

The third implication of this definition follows from not limiting what kinds of things might be judged to succeed or fail. Although we are primarily dealing with the failure of artefacts here, this definition is applicable to things other than artefacts. Students fail when they do not meet relevant success criteria such as minimum marks or minimum class attendance.

---

[2]Department of Defense, *Mil-Hdbk-338b Military Handbook: Electronic Reliability Design Handbook* (Washington: US Department of Defense, 1998). http://www.relex.com/resources/mil/338b. pdf

This book uses a single sense of the term failure but in some fields a range of apparently similar terms are used to refer to failure. One internationally used standard refers not only to failure, but also to fault and defect.[3] While such distinctions may have value in specific situations, the notion that failures can be graded is misleading, primarily because it conflates failure with the consequences of failure but also because it confuses the decision as to whether something succeeds or fails with the criteria used to judge that. For this reason, we will consider only one concept here, failure. The role of the effects of failure and revision of success criteria are discussed in more detail later.

3.3 Success Criteria

Since success exists in relation to failure, each reference to failure is also a reference to success and vice versa. This means that the term success criterion could just as validly have been failure criterion, or something similar. The term success criterion is used here for simplicity but it is important to bear in mind that it applies to both success and failure.

Success criteria may take many forms. Although the carbon microphone was a technological device, Edison also had to satisfy non-technological success criteria. Technologically, it had to transmit articulate speech, not degrade over time (the problem with Fluff) and be compatible with Western Union's existing telegraph system. It also had to satisfy non-technological criteria including not infringing existing patents, being easy to operate and a reasonable cost to manufacture.

Success criteria can change over time but also vary between people. We see this with functionally similar products. A success criterion for one purchaser might be price while for another it might be durability. The result is that an item that meets the second person's criterion for durability may fail the first person's criterion of price. Many disputes over the success of technology turn out to be not disputes over success or failure but disputes over which success criteria should be applied. Because success criteria vary between people, such disputes revolve around value judgments over the selection and prioritisation of success criteria rather than being disputes over performance against success criteria.

Taken together, the issues of who decides, what success criteria are applied and when judgments of success are made mean that decisions about success or failure are not absolute or permanent.

Chapter 2 touched on Thomas Hughes's reverse salient metaphor, crucial problems that hold back the development of a technology. In terms of success criteria, a reverse salient is the failure of an artefact to meet crucial success criteria that hold

---

[3]ASHRAE, "ANSI/ASHRAE/IESNA 90.1:2010: Energy Standard for Buildings except Low-Rise Residential Buildings," (Atlanta, GA: American Society of Heating, Refrigerating, and Air-Conditioning Engineers, Inc., 2010).

back overall progress of the technology. Edison's realisation that incandescent lamps required a high filament resistance was such an advance, high filament resistance being a crucial success criterion for an economically and technologically viable electric lighting system. Many other inventors before Edison developed workable incandescent lamps but their low filament resistance meant that they were limited to laboratory demonstrations or to lighting individual rooms or buildings where the wiring cold be kept short. These earlier incandescent lamps failed to meet the criterion of suitability for use an electric lighting utility that could compete with gas lighting. Edison succeeded in developing electric lighting because he identified a reverse salient success criterion, filament resistance, and then succeeded in meeting it.

3.4   Success Clues

While we use success criterion to judge whether something succeeds or fails, we also use something similar to identify candidates that are likely to succeed or fail. These will be referred to *success clues*. When Edison began developing the Phonograph in 1877, he used waxed paper as the recording medium but discovered that the wax chips produced during recording process interfered with sound reproduction. He then set out to find a recording medium that avoided chips, eventually using tinfoil and changing the recording process from cutting to indenting. In this instance, noise-free recording was the success criterion to be met, while not forming chips was a success clue, an aid to meeting that criterion.

Success criteria are always success clues but the reverse is not necessarily true. The distinction between the two relates to the reason for aiming to satisfy each. In the Phonograph example, Edison's primary success criterion was to record and repro-duce sound without extraneous noise. Finding something that did not produce chips was a way to that end, not an end in itself. Had he devised a recording method that used wax as the recording medium but was able to remove the wax chips as they were produced, he would have met the success criterion of reproducing sound but not the success clue of being chip free. The success of an artefact does not depend on meeting success clues.

Rules of thumb are a common form of success clue. Though apparently simple compared with rigorous analysis, rules of thumb simplify the solution of complex underdetermined problems by reducing the number of variables to be handled. Since engineering problems tend to be highly underdetermined, success clues such as rules of thumb are crucial to their solution.

While some success clues are explicit to a greater or lesser extent, others are comparative only. An example of this is the notion of simplicity, sometimes expressed as the success clue "keep it simple" (KIS). Edison did not state this explicitly when developing the carbon microphone but it is evident in the direction he took, the final design sold to Western Union being far simpler than many of his earlier devices. It is evident that Edison was aware that simplicity was a predictor of success despite it not being an explicit success criterion.

3.5 Problem Redefinition

The situational nature of success criteria is evident in many areas of artefact creation. Alexander's widely read architectural textbook observes that architects "are never capable of stating a design problem except in terms of the [failures] we have observed in past solutions to past problems".[4] This is a relevant description of other artefact creation processes including invention and has a number of consequences.

Firstly, processes aimed at creating novel artefacts like architectural designs and inventions are fundamentally failure reduction processes, that is processes aimed at reducing the chances of the artefact failing. Secondly, the creation of novel artefacts must, of necessity, begin with some uncertainty about what the problem is even if the artefact is similar to many in created in the past. If the artefact is novel to any extent, it will have a novel problem definition and the task of problem definition is one of determining relevant success criteria. This means that the process of developing novel artefacts is accompanied by a process of continually refining and redefining the problem definition by adding and modifying success criteria. Just as Edison added more and more success criteria while developing the carbon microphone, an architectural design does not spring into being as a finished drawing but develops over time as the problem is redefined and success criteria added, modified and solved.

Thirdly, and crucially, continuous problem redefinition means that those who create artefacts can never be certain they have eliminated all potential for failure. There are no perfect artefacts because we can never be sure we have found and solved all success criteria that might lead to the artefact being judged a failure.

3.6 Success Frameworks

Problem redefinition involves continuously identifying, developing and redefining success criteria and success clues. I will refer all an artefact's success criteria and success clues taken together as its success framework. It is evident from Edison's development of the carbon microphone that he was not only particularly skilful at creating successful artefacts but also skilful at identifying relevant success criteria and success clues, that is of developing an artefact's success framework. The process

---

[4]Christopher Alexander, *Notes on the Synthesis of Form* (Cambridge, Massachusetts: Harvard University Press, 1964), 102.

problem redefinition by creating a success framework in parallel with the creation a novel artefact is not unique to Edison or to architects but is common to everyone who aims to create a successful artefact.

Creating a success framework is not simply a series of decisions about of which success criteria to include or omit, it is also one of prioritising competing success criteria. An engineer designing an aircraft component will have success criteria that include strength, weight and cost, criteria that will almost certainly be in conflict since the cheapest alternative will not be the strongest or the lightest, nor the lightest, the strongest or cheapest, and so on. This conflict is normally resolved by expressing success criteria in quantitative or qualitative terms so that the combination in this example might be strong enough for the task, less than a budgeted total mass and costing less than a budgeted amount. This prioritisation can also mean that one artefact may be judged more successful than another (exceeded a success criterion by a greater margin) or perhaps less of a failure against the success criterion (failed by a smaller margin), or it might both succeed and fail against competing success criteria.

Although the test against a success criterion is a yes/no, pass/fail decision, the fact that success criteria can be varied at any time can result in the judgement of success appearing to be one of graded failure. This is observable in the acceptance of what might be described as a partial failure against a success criterion whereas in reality, the success criterion was varied so that failure became a success against the varied criterion. Varying success criteria to achieve acceptability is something we do frequently. A person goes into a shoe shop to buy a pair of shoes intending not to pay more than a certain amount. They find however that of the shoes on offer, the pair that meets their criteria of comfort, style and so on cost more than they intended but they still buy them. The shoes fail against their original success criteria but succeed against revised criteria that, in effect, have been revised so that style, comfort etc. are not compromised but the upper price limit has been increased. If the person stayed with the original maximum price and bought a pair of shoes that were so uncomfortable they only wore them once, the purchase would have been a failure.

The result of this process of revising success criteria and managing competing criteria is that success frameworks can be fluid, a consequence of prioritising success criteria against one another. In some cases, prioritising may involve varying the success criterion, say by revising its quantitative value so the criterion is more easily met. In other instances, the success criterion may be discarded. In practice, prioritising success criteria involves compromise. Sometimes the compromises may have the unintended effect of the artefact failing other success criteria that are more critical.

Since the success framework is created for a purpose (to define success of the artefact), the success framework is itself an artefact and so has its own success criteria that are different from those of the original artefact. Among these criteria are that it be comprehensive and include all success criteria applicable to its artefact. The success framework thus has its own, separate, success framework and potentially a life of its own, separate from the associated artefact.

This potential for separating the success framework from the artefact is the basis for reverse engineering (Chap. 11) in which the success framework for one artefact is used to create a new artefact, possibly with a very different physical form from the original but meeting the same success criteria and having the same functions as the original.

3.7 Identifying Failures

The development of the carbon microphone shows that Edison was not only very good at solving failures; he was also very good at identifying success criteria. Part of Edison's process of building a successful device involved identifying a wide range of success criteria. We can see him working in this way when he analysed the ways in which his early telephone designs failed to transmit articulate speech, in so doing identifying the parts of speech not adequately reproduced. As Edison had no previous experience of speech pathology, he invented his own terminology including "hissing" and "clang" to describe types of sounds. Edison and his associates tested their microphones by whispering, whistling and speaking phrases like "Physicists and Sphynxes in majestical Mists" and "The majestical myth which physicists seek".[5] Edison's notebook entries on the carbon microphone refer only to its performance when transmitting speech and not music, since his objective in developing it was the transmission of "the complicated sonorous vibrations of the human voice with all its modulations".[6] The telephone was to be used for transmitting telegraph messages and not for entertainment, so it did not need to handle the frequency range of musical instruments or satisfy the related success criteria.

Koen observes that one way engineers solve an apparently insoluble problem is to redefine it so it becomes a soluble one.[7] In a similar way, Edison redefined the task of inventing a microphone by concentrating on the limited frequency range required for speech rather than the wider range needed for music. While Legat had reported that Reis's telephon transmitted melodies "with astonishing correctness", Edison chose not to emulate Reis's success.[8] Instead, he concentrated on parts where Reis had failed, discarding a success criterion that Reis's device met and redefining the problem to the more limited success criterion of transmitting only speech.

---

[5]TAED NV12:12.

[6]TAED TI1:38.

[7]Billy Vaughn Koen, *Discussion of the Method: Conducting the Engineer's Approach to Problem Solving* (New York; Oxford: Oxford University Press, 2003), 89.

[8]Legat, "Reproducing Sounds on Extra Galvanic Way".

3.8   Identifying Success Criteria

For something to be judged a success, it must succeed against all success criteria whereas to be judged a failure, it need fail only one. This was a crucial problem for Edison with the carbon microphone. While it satisfied all other success criteria, within a few years it became apparent that it failed the most important success criterion set by Western Union: circumventing Bell's telephone patent. Faced with this failure, Western Union took a pragmatic path and sold its telephone business at a profit to Bell Telephone, completing the process that began with Bell offering the telephone to Western Union in 1876 and Western Union outbidding Bell for the rights to the carbon microphone in 1878.

Edison's first telephone sketches in July 1876 indicate that his attention was confined overcoming the technical failures he had identified in the Reis and Bell telephones. As development progressed over the next 3 years the range and detail of his success criteria broadened. Initially he was pleased to get any sounds, writing in his laboratory notebook about an early device that at least "you could tell that someone was talking and if you knew what they were saying it sounded awful like what they were saying".[9] As the device developed, Edison added more success criteria and clarified existing. In addition to rapidly generating new ideas for solutions, Edison also continuously sought new success criteria since the greater the number and range of success criteria met, the more successful his invention would be.

3.9   Identifying Limits

The process of identifying success criteria by analysing failure highlights another way in which failure can yield knowledge that is not available from success. In understanding a failure, identifying the relevant success criterion it does not meet is a critical step to overcoming it. The valuable knowledge gained, the knowledge that can be used to prevent future failures and to improve artefacts, comes from identifying the specific criterion that the artefact failed to fulfil and the circumstances under which failure occurred. Until failure occurs and the detail of the success criterion identified, all we know is that failure is possible at some value or situation more extreme than the artefact has so far experienced. The problem is that we cannot know whether the margin before failure is 50% or 500%. It is hard to get this kind of information from analysis of success.

We can see Edison seeking limiting values for success criteria when he tested different stiffness springs in the carbon microphone in March 1878, eventually settling on a rigid brass tube. We can also see him similarly seeking to establish the successful range of materials when he tested various Fluff mixtures, discovering

---

[9]TAEB 3:767n1.

that even pure plumbago, without silk fibres, worked. While both these limit-testing exercises yielded an answer at the extreme, the most common result of such testing (and the kind that Edison was seeking in these examples), is to determine a limit that lies on a continuum rather than at its extremes. By defining a limit for failure, the safe margin for success over failure can be reduced and weighed against other factors. In practice, approaching limits means inventors and other technologists work not with absolutes, but with probabilities, that is, with a very small but still finite risk of failure. The key to success is to approach, but not exceed, the circumstance at which failure occurs.

3.10 Everyday Use of Failure

This book focuses on Edison's use of failure approach in inventing but the use of failure is not exclusive to inventors like Edison. It is a small step to identify similar examples in disciplines such as architecture and software development. Indeed, the concepts developed can be extended to approaches used by anyone who seeks to create any kind of novel artefact. All who seek to create novel artefacts face similar problems including identifying relevant success criteria, problem definition, knowing when to stop, dealing with an absence of relevant theory and working with a large number of unknowns. They must deal with many possible success criteria, each of which can be given different emphasis or excluded from consideration. This means that there is no single solution or even a small number of solutions. One might visualise a composer or novelist dealing with similar problems. The writer of a detective story may have an overall success criterion (a successful, published book) but needs to weave clues (and red herrings) into the story for it to progress and keep the reader's interest, while coming to a believable conclusion.

A common conversation between people discussing doing something new runs like this:

Conversation	Interpretation
"If we do this, X will happen."	*A potential failure is identified by testing against success criteria*
"Oh. In that case we should do Y"	*Solution proposed to meet the success criterion just identified.*
"But if we do Y, we will avoid X, but run the risk of Z"	*Another potential failure is identified by testing against other success criteria.*
and so on...	

Such dialogs continue as a sequence of suggestions, tested against established success criteria and modified or abandoned in order to avoid potential failure. Such dialogs also go on in the minds of individuals – I do it as I write this. The success criteria applied are not limited to the technical issues but can include anything relevant like budget, time to design or that is possible to build. For people building a house in the wet tropics, putting the roof up first will keep out the rain while the rest is built, but what will hold up the roof? The more testing of this kind that is done, the greater our chance of producing a successful artefact. What is notable, is that it is testing in the mind and in dialogs.

Such processes can produce success out of an awareness of failure, but they are not without hazards. One hazard is that failure can be pursued to the point that the fear of failure results in doing nothing. This is particularly the case when trying to prevent unlikely forms of failure (low risk events). A more common hazard is over-compensation when trying to avoid potential failure. This is commonly referred to as overdesign and tends to be more prevalent than the opposite (underdesign) because while underdesign becomes obvious through failure, overdesign may remain hidden for the life of the artefact. If this persists on a wide enough scale, it is possible for overdesign to become institutionalised, the norm or "accepted practice", simply because it does not fail.

3.11 Failure as Paradox

It is a paradox that, despite the crucial role of failure in creating novel artefacts, successful artefacts rarely reveal the role that failure played in their creation. Occasionally we can detect the effect of past failures from features of an artefact. When we encounter an artefact, say an aircraft, with duplicated systems when only one would serve, a reasonable inference is that in the past this kind of artefact failed because its single system failed. Mostly however the failures of artefact's creation are hidden, perhaps forgotten even by the artefact's creators. The invisibility of failure in the finished invention may explain the common sense view that failure is only of value in telling us what not to do next time.

A second paradox of the role of failure in creating novel artefacts like inventions is that the more thoroughly the creators of artefacts seek to make them fail, the more likely they are to succeed. While failures were significant to Edison for their potential to identify new phenomena, their primary value was to identify potential problems to be overcome in order to improve the invention. Edison's incentive in pursuing failures was to encounter and overcome them in the laboratory so they would not occur in the field. To do this he tested his inventions in more and more demanding situations and against more and more success criteria. For Edison, finding and overcoming more failures was the path to more successful inventions. While Edison's success as an inventor is well established, the way he produced successful inventions is better understood as one driven by the search for and exploitation of failures, than of a search for successes.

The quotation from Bill Gates discussed in Chap. 5 implies that our society should aim for continuous innovation. However, this is a society that is increasingly risk-averse. If innovation is pursued – and it is more than token innovation – frequent failure will be the result. There is thus a tension between a desire for innovation and risk-aversion, so it is worth asking what alternatives there might be to innovation. This is an important question but it is usually only answered rhetorically by advocates of innovation, and then in the negative, typically with the assertion that not innovating means stagnating.

The analysis in in this chapter suggests at least one alternative. Innovation is a relative term: for something to innovative part of it must be new. However, only part of it can be new since innovation must build on previous successful innovations. Even Edison's revolutionary Phonograph (Chap. 7) incorporated the centuries old technologies of the screw and crank. Chapter 4 argues that the risk of failure of innovation increases both with increasing complexity and with the amount that is new. Innovators can increase their chances of success and reduce the risk of failure by aiming for simplicity rather than complexity, and by making as much use as possible of mature technologies for which success criteria are well established.

Neither of these approaches is radical, indeed the second might be described as craft or craftsmanship, excelling at what is known to be successful and doing it better. This suggests that the advantages claimed rhetorically for innovation might be achieved at lower risk by emphasising simplicity and skilled use of mature technologies rather than by prioritising the new.

Chapter 4
Innovation and Systems

4.1 Systems

Historian of technology Thomas Hughes argues that global electric power networks
are systems in which technological, social, political and economic aspects are all
components. Hughes describes a system as follows:

> A system is constituted of related parts or components. These components are connected by
> a network, or structure, which for the student of systems may be of more interest than the
> components. The interconnected components of technical systems are often centrally con-
> trolled, and usually the limits of the system are established by the extent of this control.
> Controls are exercised in order to optimise the system's performance and to direct the system
> toward the achievement of goals. The goal of an electric production system, for example, is
> to transform available energy supply, or input, into desired output, or demand. Because the
> components are related by the network of interconnections, the state, or activity, of one
> component influences the state, or activity, of other components in the system. The network
> provides a distinctive configuration for the system.[1]

The key features of Hughes's concept of a system are:

- The system is composed of components.
- Components are connected or related by a network or structure.
- Interconnections between components mean that changing one component influ-
  ences other components in the system.
- The system is often centrally controlled.
- The limits of this control determine the system's boundaries.
- Control is exercised to achieve goals for the system.

These features can be illustrated by their role in an electrical power utility:

---

[1]Thomas P Hughes, *Networks of Power: Electrification in Western Society 1880–1930* (Baltimore:
Johns Hopkins University Press, 1983), 5.

© Springer Nature Switzerland AG 2019   61
I. Wills, *Thomas Edison: Success and Innovation through Failure*, Studies in History
and Philosophy of Science 52, https://doi.org/10.1007/978-3-030-29940-8_4

- Components: Incandescent lamps, cables and generators are components but so are non-technical components. Human actors like Edison and his financiers are components, as is the financier's capital.
- Network: The technological components of an electric power system are connected by a structure or network via generators, cables, transformers, and so on. Non-technological components can also form networks. In the early 1900s, Samuel Insull, who had been Edison's secretary, built a financial network linking numerous small Edison electric companies under the Commonwealth Edison umbrella. In so doing, Insull invented the holding company, a new business entity and a system.[2]
- Interconnection and interaction: Components are dependent on, and related to, one another. This is most obvious with technological components. Connecting a large power load to the extremity of an electric power system is reflected in increased current in the cables and increased load on the central generating plant. Non-technological components also interact. Edison's analysis to determine the required filament resistance for his incandescent lamp reflected the interaction between a non-technological component, the price of copper for cables, and a technological component, the current the cables were to carry. Regardless of its size, a system derives some of its characteristics from interactions between its components, characteristics that are not the result of the components acting alone.
- Control: Control is the key to understanding this type of system because, Hughes argues, systems are changed in extent (expanded) to increase control. For Hughes, something not controlled by a system is part of the system's environment but if brought under control, it becomes a component of the system, interacting with other system components. Expansion of control drives system expansion.
- Boundaries: New components may be added to an existing system to increase control and networks expanded to increase centralised control. System expansion is not limited to technological components. An electric utility that buys a coal mine to supply its power station makes the mine a component of its system and expands the system boundary. The cost of coal changes from an interaction between the system and its environment, into an interaction between components within the system. On the technological level, electrical power systems are controlled to maintain voltages and to balance loads. The need to balance loads, a technological objective, led early electrical utilities to promote a non-technological objective, the use of electric domestic appliances because these provided load during the day to balance night time lighting load. In Hughes's control-oriented systems, the emphasis on control means that they tend to expand rather than contract.

---

[2]Samuel Insull (1859–1938) English-born financier. Insull became Edison's secretary in 1881 but later moved to Chicago to become president of Chicago Edison. By progressively merging with smaller electric utility companies Insull built the Commonwealth Edison Co into an electric utility empire. When his utility holding company collapsed during the Great Depression he was prosecuted for mail fraud and antitrust offences but later acquitted. Insull is credited with creation of electric utilities as a network and of the concept of the holding company.

- Goals: The goals of the system are many. Edison's earliest goal was simply to turn the electrical energy into light. Over time his technical goals expanded but he also added non-technical goals particularly that of making electric lighting commercially competitive with gas lighting. To achieve this goal Edison analysed the economics of his system, concluding that the economic viability of electric lighting depended on reducing the amount of copper in the cables and to achieve this he needed high resistance lamps. Edison's initial objective of producing an incandescent lamp expanded to include the high resistance lamp, expanding further to include a generator to match the high lamp resistance since the existing generators for low resistance arc lighting were unsuitable. Eventually Edison expanded the scope of his electric lighting system patents to cover a broad range of components. Non-technological goals also shape such systems. Noting the differences between the structure of electrical distribution networks of London and Berlin in 1914, Hughes argues that their structure reflected the political values of their respective governments. In Britain, an emphasis on diversified local government resulted in many small power stations, while the centralised government in Germany led to a few large stations.[3]

4.2 Functional Systems

Hughes's notion of a system is appropriate to large technological systems like electric lighting but its emphasis on control limits its usefulness in other situations where control has either a minor role or no role at all. (Chapter 9 looks at Edison as a builder of this kind of large technological systems.)

This chapter introduces a different system-based approach, referred to as *functional systems*. A functional system is simply a collection of components that interact and are related by a structure. A functional system is similar to a control-oriented system in that it involves components that interact within a structure. Unlike a control-oriented system the objective of which is control, the objective of functional system is to serve its functions, whatever these might be. The advantage of functional systems is that they are flexible: their boundaries, composition and relationships are not fixed or limited so, unlike control-oriented systems, their boundaries can expand or contract as needed.

Functional systems can accommodate Hughes's control-oriented systems by treating control as a function. Hughes's notes that, "Controls are exercised in order to optimise the system's performance and to direct the system toward the achievement of goals."[4] For Hughes, control is exercised to achieve the goal of the system.

---

[3]Hughes, *Networks of Power: Electrification in Western Society 1880–1930*, 15–17.
[4]Ibid., 5.

In functional systems, control is just one of many possible means by which the system's functions or goals are achieved.

4.3   Systems, Components and Interactions

A consequence of treating systems as a collection of components that interact within a structure is that some of the properties of a system come from interactions between components rather than from the components themselves. For these interaction-derived properties, component properties are significant only in the way they affect interactions. That is, some component properties are significant only in relation to interactions. This means that it is possible to get the same system properties using different components with or without different properties, provided the combination of properties in interacting components results in the same interactions. While such substitution is feasible in principle, in practice, as we will see later in this chapter, even very simple systems like Edison's early cylinder Phonograph involve a large number of interactions. The result is that the likelihood of substituting components with new properties and getting the same system properties is extremely low because even if we succeeded in getting an identical interaction between two components, the substitute components would interact differently with other components in the system, altering the properties and behaviour of the system as a whole.

4.4   Systems, Functions and Means

An artefact is something intentionally made or produced for a purpose. It follows that artefacts can be treated as consisting of two aspects: the created part of the artefact, and its purpose or functions. Using Edison's carbon microphone as an example, the microphone is the means and converting sound to varying electrical resistance its function. Put simply, an artefact consists of the functions it achieves and the means by which they are achieved.

Functional systems were defined earlier as a collection of components that interact in a structure to serve a function or functions. There is thus a parallel between artefacts and functional systems: both consist of something intended to serve a function or functions. This parallel can be expressed as follows:

Invention	=	Functions	+	The means by which they are achieved.
Functional system	=	Functions	+	A collection of components that interact and are related by a structure to achieve the system's functions.

As with the earlier discussion of functions in relation to artefacts, the functions of a functional system separate from the functional system. There are a number of consequences of the parallel between artefacts and functional systems. Firstly, we can treat the means part of inventions as systems, making functional systems a valuable analytical tool for understanding invention and novelty. Secondly, the means of an invention can be treated as a system; a collection of components that interact and are related by a structure to achieve its functions. Thirdly, as we saw earlier through Edison's development of the carbon microphone, the creation of an artefact, in that case a patentable invention requires the building of a set of success criteria for the artefact (its success framework) related to it successfully fulfilling its functions. The success framework can thus also be treated as a system.

A common form of patent is one that claims novelty through an improvement to an existing technology. In all, Edison was awarded 388 electric lighting patents, the first of which begins:

"Be it known that I, Thomas A Edison of Menlo Park, in the State of New Jersey, have invented an Improvement in Electric Lights, of which the following is a specification."[5]

Edison then proceeds to explain how his invention, a means for controlling filament temperature (a new success criterion he has identified), improves the performance of existing platinum filament incandescent lamps. In this patent, Edison achieved novelty by adding a new function (temperature control) and success criterion (the filament should not overheat and melt).

Later in the patent, Edison claims that he has tested, "a large number" of devices for achieving this function but the one for which the patent is sought is the "most convenient". In doing this he links the primary, novel, function of filament temperature control to an implied function of convenience, the primary function being possible, Edison implies, by many means but only the one proposed for the patent achieves both primary and implied functions.

One function or set of functions can potentially be met by many different means. Since the means of an invention is a functional system, it implies that its functions can be met by many different functional systems. Changing the artefact by adding more functions or changing or adding success criteria, or by changing the components, interactions or structure of the artefact/functional system potentially creates a new functional system and a new artefact. A common way inventors produce novel (and therefore patentable) inventions is by devising new means for achieving existing functions. They can also create inventions by devising new ways of using existing means to achieve new functions, or they can do both in one invention. Restating this in terms of functional systems, we can say that inventors devise new functional systems to achieve existing functions, or devise new functions for existing functional systems or devise new functional systems for new functions.

Control-oriented systems like electrical utilities tend to increase in size and complexity and consequently their boundaries tend to expand. In contrast, functional systems have no fixed boundaries and can be changed in size to suit the analysis. Since a functional system exists to serve its functions, the functional system's

[5]Thomas A Edison. Electric Lights. US Patent 214,636, filed 14 October 1878, and issued 22 April 1879.

boundaries can expand, contract or move to match changes in these functions and components.

We can illustrate this by considering Edison's incandescent lamp because in functional system terms it is both a component of a larger electric lighting system and a functional system in itself. Viewed in isolation, as a functional system, its function is to convert electricity to light. We can also move the boundary for functional system analysis to isolate parts of the lamp, revealing other functional systems. The lamp filament in a vacuum is a functional system with one set of functions, primarily to convert electrical energy to light. Putting a boundary around Edison Screw base and lamp holder creates another functional system the primary function of which convenient connection to electrical wiring. Moving the boundary again, we can also treat a group of lamps wired in parallel as another functional system a function of which is to permit single lamps to be switched on and off without affecting the whole system. Unlike the control-oriented model of systems, there is no single functional system but many.

Not only can we narrow the boundaries of a functional system to reflect a different function (lamp, lamp plus lamp holder or just filament) but also broaden boundaries by adding functions. We can draw the boundary wider around Edison's lamp to include not just the lamp itself but Edison's high vacuum pump and his Menlo Park laboratory each of which were components that interacted to achieve different functions. The interactions between the lamp and vacuum pump relate to the function of achieving a longer lamp life. The functional system that includes Edison's lamp, vacuum pump and Menlo Park is aimed at another set of functions, one of which was to generate ideas for inventions.

We can also treat people as components, each with their own properties, interacting together to create something separate so that we can view Edison and Charles Batchelor as a functional system in which Batchelor's steady, meticulous approach interacted productively Edison's imaginative but somewhat flighty ways. The function of the Edison-Batchelor system was to produce valuable and novel inventions. Remove Batchelor, and Edison would undoubtedly have produced fewer inventions. Remove Edison, and Batchelor might have produced none.

Still more functional systems can be identified at Menlo Park by including not only the people and inventions but the laboratory's physical resources, such as its large store of chemicals and technical library. Yet another Menlo Park boundary produces a functional system that includes the newspaper reporters who provided the public with a ready stream of stories about Menlo Park, Edison and his inventions. In this system, Edison interacts with the reporters who interact with the public via their newspapers. Among the functions of this system are selling newspapers, entertaining the public, building Edison's reputation and attracting finance to his projects.

As we expand the boundaries of the Menlo Park functional system, adding more technological and non-technological components we move from the simple functional system of a few lamp components or a few people to a large technological system. It is also an expansion that can extend in time. Edison's high resistance carbon filament lamp became a component of his direct current electric lighting system but it also became a component of Tesla's alternating current system because alternating current involved the same interaction between lamp resistance and the cost of copper in cables. Later, when commercial tungsten filament lamps were introduced, they became plug-in replacements for Edison's carbon filament lamps. The filament retained the same crucial property despite a significant difference in its material.

Although the concept of a functional system has been introduced in this chapter, the reasoning involved is not new. In his account of the failure of Aramis, an ambitious public transport system for Paris, Bruno Latour treats people, engineering components and physical laws as components (actors) in what are essentially functional systems.[6] At the end of his account, Latour's semi-fictional protagonist explains the failure of the project by concluding that each of the key actors saw Aramis differently. In effect Latour argues that Aramis failed because each treated it as a different functional system.

4.5   The Phonograph as a Functional System

The Phonograph was the invention that cemented Edison's reputation as the creator of not just novel but revolutionary inventions. We can now turn to the use of functional systems to analyse innovation in invention, using Edison's early Phonograph as an example (Fig. 4.1).

To record sound with this Phonograph, Edison wrapped a piece of tinfoil (the recording medium) around the grooved horizontal cylinder. This was fixed to a long screw, the screw and cylinder being rotated by the crank in Edison's hand in the photograph. The person whose voice was being recorded shouted into the mouthpiece, causing the diaphragm to vibrate. An inscribing point attached to the diaphragm then made small dents in the tinfoil in response to the sound as the foil moved past. To reproduce the recorded sound, the cylinder was returned to its starting position, the inscribing point again brought into contact with the tinfoil and the handle cranked. As it moved past the inscribing point, the dents in the tinfoil caused the inscribing point to vibrate and the connected diaphragm to emit the mechanical vibrations as sound.

Despite its revolutionary impact, this Phonograph is remarkably simple and can be reduced to a handful of components (Fig. 4.2): the tinfoil recording medium, the inscribing point, the diaphragm and something to hold the parts together.

The first step in analysing this as a functional system is to decide at where to place its boundaries. One possibility is to place it around the hardware of the Phonograph but this hardware alone cannot perform the functions we expect of a Phonograph because this system can neither record nor reproduce sound. To achieve these functions we need to expand the functional system boundary by adding a means to

---

[6]Bruno Latour, *Aramis, or, the Love of Technology* (Cambridge, Massachusetts: Harvard University Press, 1996).

Fig. 4.1 Thomas Edison and his Phonograph, photographed in Washington, April 1878 (Mathew Brady, "Thomas Edison, Full-Length Portrait, Seated, Facing Front, with Phonograph," (Washington: Library of Congress, 1878). http://www.loc.gov/pictures/item/89714876/)

Fig. 4.2 Simplified schematic drawing of the Phonograph in Fig. 4.1

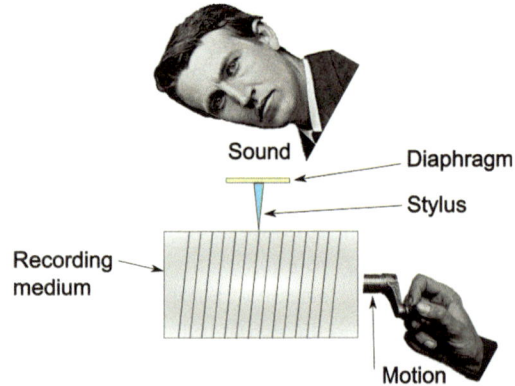

move the tinfoil relative to the recording and reproducing point and a sound source to be recorded. To reproduce the recorded sound we again need the hardware and motion plus a person to hear it.

Edison described the Phonograph in similar terms in his first Phonograph patent:

> The invention consists in arranging a plate, diaphragm, or other flexible body capable of being vibrated by the human voice or other sounds, in conjunction with a material capable of

Fig. 4.3 The Phonograph
as a functional system.
Reading from top to bottom,
it records sounds while
reading from bottom to top it
replays them

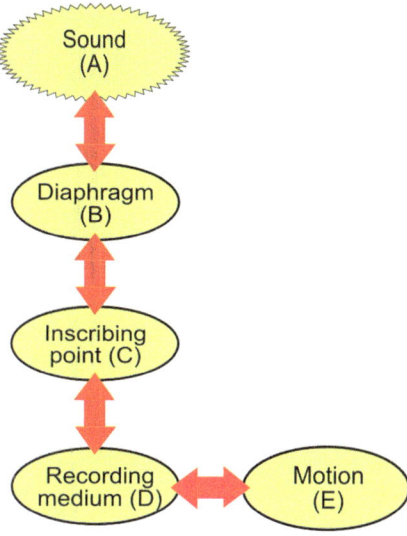

registering the movements of such vibrating body by embossing or indenting or altering such material in such a manner that such register marks will be sufficient to cause a second vibrating plate or body to be set in motion by them, and thus reproduce the motions of the first vibrating body.[7]

Figure 4.3 shows these components schematically as a functional system.

Before Edison settled on a cylinder configuration for the Phonograph he tested many variations seeking to understand the ways in which they failed. One such failure with waxed tape versions was the difficulty human operators had in maintaining constant speed during recording and replay. To deal with this, he proposed replacing the operator with a clockwork drive to move the recording tape (see Chap. 7) but with the cylinder Phonograph, he adopted the simpler solution of a heavy flywheel seen on the far left of the Phonograph in Fig. 4.1.

If we approach the Phonograph as a control-oriented system, the clockwork drive and flywheel expand the system to increase control, in this case to control recording medium speed. As a functional system this is interpreted as Edison adding another success criterion, constant speed. Edison identified this success criterion as a consequence of a problematic interaction between components, in this instance between a non-technical component, the operator, and a technical component, the recording medium.

In both functional systems and control-oriented systems changing or adding components affects interactions with other components. Much of Edison's trial and error development work was aimed at refining such interactions because apparently minor changes to components and interactions between components could

---

[7]Thomas A Edison. Phonograph or Speaking Machine. US Patent 200,521, filed 24 December 1877, and issued 19 February 1878.

radically alter the behaviour of the invention as a whole. The success of a system as a whole depends, to a significant extent, on interactions between components.

4.6   Even Simple Functional Systems Are Complex

Perhaps the most striking aspect of the early Phonographs, apart from what they did was that they were so simple. Chapter 7 discusses the early development of the Phonograph and concludes that despite the Phonograph's simple operating principle, producing a successful Phonograph was not simple.

Figure 4.3 shows the Phonograph as a functional system of the five components identified by the letters A, B, C, D and E. These five adjacent components have four interactions with each other, shown as arrows. In a functional system, the boundaries can be moved to redefine a new functional system so this five component functional system also contains a number of other systems. Easily identified subsystems are the diaphragm and inscribing point BC; the inscribing point and recording medium CD; and the human operator AE. We can continue this approach with three and four component subsystems giving the following components and systems:

- Five single components: A, B, C, D, E
- Four two-component subsystems: AB, BC, CD, DE
- Three three-component subsystems: ABC, BCD, CDE
- Two four-component subsystems: ABCD, BCDE

Not only can components interact with each other (for example the inscribing point C with the recording medium D) but components can interact with subsystems and subsystems with other subsystems (for example motion E with the inscribing point-recording medium subsystem CD). The possible interactions for a five component system are shown in Table 4.1.

In Table 4.1, P indicates a possible interaction and 0 an invalid interaction. For example C cannot interact with BCD because it is contained within this subsystem. The cells in the upper right half are blank because they duplicate the values on the lower left half (interaction BD is the same as DB). Summing the possible interactions gives a total of 35, significantly more than the number of components. We can repeat this for systems with other numbers of components. For example a three component system yields five possible interactions; four components, 15 interactions; five components 35 interactions; six components, 71 interactions; seven components, 122 interactions, and so on. These interactions are not dependent on the nature of the system or its individual components; they are the consequence solely of these components being part of a system. That is, the analysis in Table 4.1 would apply to any five component system.

This indicates that the number of interactions exceeds the number of components by a considerable margin and increases very rapidly as the number of components increases. A system with 100 components has almost 3 million possible interactions. Although in this analysis of the Phonograph all possible interactions are valid, in

Table 4.1 Possible interactions for a five component system

	A	B	C	D	E	AB	BC	CD	DE	ABC	BCD	CDE	ABCD	BCDE
A	0													
B	P	0												
C	P	P	0											
D	P	P	P	0										
E	P	P	P	P	0									
AB	0	0	P	P	P	0								
BC	P	0	0	P	P	0	0							
CD	P	P	0	0	P	P	0	0						
DE	P	P	P	0	0	P	P	0	0					
ABC	0	0	0	P	P	0	0	0	P	0				
BCD	P	0	0	0	P	0	0	0	0	0	0			
CDE	P	0	0	0	0	P	0	0	0	0	0	0		
ABCD	0	0	0	0	P	0	0	0	0	0	0	0	0	
BCDE	P	0	P	0	0	0	0	0	0	0	0	0	0	0

practice some interactions will be inconsequential while others will be critical to the success or failure of the artefact. The problem in developing an innovative artefact is that initially we cannot be sure which interactions are inconsequential and which are critical. Gooding's observation, that, "with respect to novelty, *everyone* is a novice" is a relevant caution here.[8] With many unexplored interactions, there are many opportunities to be a novice.

4.7 Using Functional Systems to Identify Novelty and Innovation

Each of the Phonograph's physical components was familiar to Edison's audience and the Phonograph's method of operation so simple that it could be understood even by people with no technical expertise. If we attempted to identify the Phonograph's novelty only by examining its components, we might conclude that Edison invented nothing, that he just assembled a few familiar components, most of which, like the crank and screw, had been in use for thousands of years. Indeed, the Phonograph is so simple that it could, in principle, have been invented centuries earlier. It is far simpler and requires less precision than, for example, than de Dondi's fourteenth century clockwork planetarium.[9] The only component not readily

---

[8]David Gooding, "How Do Scientists Reach Agreement About Novel Observations?," *Studies in History and Philosophy of Science Part A* 17, no. 2 (1986): 208.

[9]Giovanni de Dondi, *The Planetarium of Giovanni De Dondi Citizen of Padua: A Manuscript of 1397*, trans. G.H. Baillie and H. Alan Lloyd, Monograph (Antiquarian Horological Society); (London: Antiquarian Horological Society, 1974).

available before the Industrial Revolution was tinfoil although, as Edison and Bell showed, recordings could be made successfully on wax.

The answer to this puzzle is that even though the individual components were well known, Edison's assembly of them as a functional system was novel. It is true that Edison's Phonograph resembles Scott de Martinville's 1857 Phonautograph in its mode of recording.[10] However Scott de Martinville's used a bristle to record sound as marks on a soot-covered drum whereas Edison first used a cutting point on waxed paper then an inscribing point on tinfoil. The basic components may have been similar but Edison added an interaction between components not present in the phonautograph: an impression in the recording medium that was robust enough to move the reproduction point to vibrate the attached diaphragm and produce sound. Edison's innovation was to add a new function, reproduction of recorded sound and the means to achieve it through an interaction between components, replacing Scott de Martinville's soot covered recording surface with one robust enough to permit reproducing the recorded sound. Edison identified a new potential interaction then invented means to achieve it.

A similar problem with identifying novelty is evident in the incandescent lamp. Friedel and Israel list no fewer than 20 inventors who produced 28 separate incandescent lamp designs before Edison dating back to 1838.[11] Most of these lamps used the same components as Edison's, 90% having carbon filaments and 80% a combination of a carbon filament in vacuum. Indeed, electrically produced incandescence has an even longer history, Sir Humphry Davey having observed in 1812 that a platinum wire connected to his massive battery "instantly became red hot, then white hot, the brilliancy of the light was soon insupportable to the eye" while in 1814, another English researcher, George Singer, performed a similar experiment but with thin platinum wire in a vacuum where the wire attained a "glowing white heat".[12,13]

---

[10]Édouard-Léon Scott de Martinville (1817–1879) French printer and inventor. In 1857 he demonstrated his Phonautograph, a device that recorded sound on a soot covered drum for later inspection but was not able to reproduce the recorded sound. Joseph Henry used Scott de Martinville's Phonautograph to experiment on the nature of sound, while Alexander Graham Bell tried using it to teach the deaf to speak. Edison may also have known of the phonautograph before developing the Phonograph because he subsequently used one in July 1878 to investigate noise from New York elevated railroad. In 2008, researchers at Lawrence Berkeley National Laboratory in Berkeley, California succeeded in reproducing a short recording of, "Au Clair de la Lune" made in 1860 on a phonautograph making this the oldest known recorded sound. New York Times, "Researchers Play Tune Recorded before Edison," *New York Times*, no. 27 March (2008).

[11]Robert Friedel, Paul Israel, and Bernard S Finn, *Edison's Electric Light: Biography of an Invention* (New Brunswick, New Jersey: Rutgers University Press, 1987), 115.

[12]Humphry Davy, *Elements of Chemical Philosophy Part I Vol I* (Philadelphia: Bradford and Inskeep, 1812), 85. http://tinyurl.com/pg33dfr

[13]Quoted in Michael B. Schiffer, *Power Struggles: Scientific Authority and the Creation of Practical Electricity before Edison* (Cambridge, Mass.: The MIT Press, 2008), 17.

The English inventor Joseph Swan is sometimes claimed to be the inventor of the incandescent lamp.[14] Like Edison, he invented a lamp using a carbon filament lamp in a vacuum. In Friedel and Israel's list, Swan is one place ahead of Edison chronologically but both Swan and Edison are preceded by many earlier inventors and similar incandescent lamp designs.

Edison's lamp with its carbon filament in a vacuum was not novel in terms of its components but was novel when understood as a functional system. Edison's was the first lamp that was commercially viable, invented to integrate into a central station electrical power system. Crucial to this integration was Edison's introduction of the high resistance filament.

Edison described the novelty of this lamp in his tenth incandescent lamp patent:

> The invention consists in a light giving body of carbon wire or sheets coiled or arranged in such a manner as to offer great resistance to the passage of the electric current and at the same time present but a slight surface from which the radiation can take place.

> The invention further consists in placing such a burner of great resistance in a nearly perfect vacuum, to prevent oxidation and injury to the conductor by the atmosphere. The current is conducted to the vacuum bulb through platina wires sealed into the glass. The invention further consists in the method of manufacturing carbon conductors of high resistance, so as to be suitable for giving light by incandescence and in the manner of securing perfect contact between the metallic conductors or leading wires and the carbon conductor.[15]

We can draw from this the components of Edison's novel functional system:

- "a light giving body of carbon"
- "a nearly perfect vacuum"
- "platina wires sealed into the glass"
- "the method of manufacturing carbon conductors"
- "the manner of securing perfect contact between the metallic conductors or leading wires and the carbon conductor"
- "great resistance to the passage of the electric current"

The critical feature Edison added to the lamp functional system, and which sets his lamp apart, is not a component, but a property of one of the components, "great resistance to the passage of the electric current". It is this high resistance that permitted Edison's lamps to be connected to remote power stations via cables of economic size. Earlier lamps such as Swan's, with low resistance filaments required the generators close to the lamps to reduce the high cost of heavy connecting cables. If supplied by remote central station generators they would have required large and consequently uneconomic cables. High resistance is critical because it affects the

---

[14]Joseph Swan (1828–1914) British pharmacist and inventor. Swan invented several incandescent lamps, demonstrating one in 1860 and receiving a British patent in 1878, the year before Edison's US patent. Unlike Edison's, Swan's lamp used a low resistance filament and so were incompatible with Edison's electric lighting system. In Britain, Edison's lamps were marketed under the name Ediswan, a combination of Edison's and Swan's surnames.

[15]Thomas A Edison. *Electric Lamp*. US Patent 223,898, filed 4 November 1879, and issued 27 January 1880.

way in which the lamp interacts with other components, most importantly, cables and generators. That is, Edison's innovation was to recognise an interaction not seen by his predecessors and from it to identify a critical property that resulted in a change in the interaction between the lamp and rest of the system.

Since Edison was not the only inventor to produce an incandescent lamp and it was clearly a very lucrative invention, Edison's patent application was followed by a legal battle over priority. The judge who decided in favour of Edison emphasised the significance of Edison's introduction of high filament resistance, concluding that, "But for this discovery [high filament resistance] electric lighting would never have become a factor. It is undoubtedly the great discovery in the art of practical lighting by electricity."[16]

4.8   Functional Systems and Innovation: Newcomen's Engine

The use of the functional systems approach to identifying innovations can further illustrated by applying it to a very different invention, Newcomen's atmospheric steam engine.

The basic operating principle in Newcomen's engine is to be found in Torricelli's 1643 discovery that he could create a vacuum in a glass tube above a column of mercury. In 1673, Huygens sketched a device intended to produce motion from such a vacuum using the explosion of gunpowder in a cylinder.[17]

Huygens's simplified diagram of this concept (Fig. 4.4) shows a cylinder B containing a piston D which is attached to a weight G via a string that runs a pulley, H. In this diagram, E and F are non-return valves that allow gases in the cylinder to escape but not re-enter. To operate the engine, a quantity of gunpowder is introduced at C and then ignited; the explosion driving out the air in the cylinder through the valves E and F. Huygens believed that expelling the air would create a vacuum in the cylinder so atmospheric pressure would drive the piston down producing useful work by raising the weight G.

It appears Huygens did not build the engine but when Denis Papin attempted to build one while at the University of Marburg he discovered that Huygens's concept was fatally flawed. The products of combustion of the gunpowder greatly exceed the volume of air expelled, so no vacuum can be created. While Huygens's apparatus did not work, Papin successfully adapted the concept by replacing gunpowder with steam.

---

[16]Quoted in Conot, *A Streak of Luck*, 286.

[17]Graham Hollister-Short, "The Formation of Knowledge Concerning Atmospheric Pressure and Steam Power in Europe from Aleotti (1589) to Papin (1690)," *History of Technology* 25 (2004).

Fig. 4.4 Huygens's
proposal for a gunpowder
powered engine (Graham
Hollister-Short, "The
Formation of Knowledge
Concerning Atmospheric
Pressure and Steam Power
in Europe from Aleotti
(1589) to Papin (1690),"
History of Technology
25 (2004))

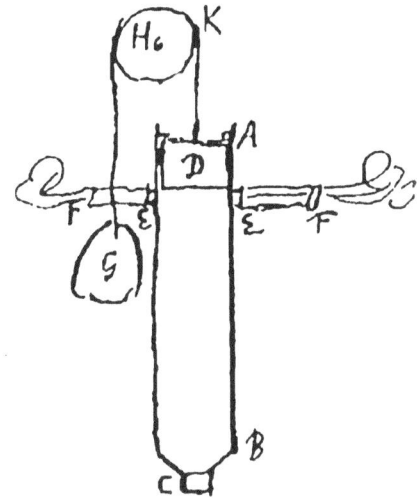

Papin published a description of the operation of his steam apparatus (Fig. 4.5) in 1690.[18] Like the device sketched by Huygens, it consists of a cylinder and piston (A and B) and rope L running over pulleys T. The apparatus was prepared by placing a small amount of water in the bottom of the cylinder. The piston B was then pressed down expelling all air until water escaped through a hole in the piston at the bottom of rod M. The hole was then plugged airtight with rod M.

Next heat was applied to the bottom of the cylinder until the water boiled, causing the piston to rise. Near the top of the piston travel, under spring pressure, the lever E engaged with the notch in the piston rod H, locking the piston in this position. The heat source was then removed and the cylinder cooled by applying water to its exterior, condensing the steam to form a partial vacuum under the piston. When the lever E disengaged from the piston rod atmospheric pressure drove the piston down with considerable force. According to Papin, a device with a piston 65 mm diameter lifted 27 kg via the pulleys T and rope L, quite a respectable performance given the theoretical maximum for this piston size is about 42 kg. In his memoir, Papin proposed producing continuous power by sequentially heating and cooling several such cylinders, the linear force being used to create a mechanical rower to propel a boat.

What was novel in Papin's apparatus? As with the Phonograph, if we consider only its components the answer will be nothing because all the components were in common use and, like the Phonograph, had been known for millennia. The piston and cylinder were common in pumps and the idea of heating the cylinder common in cooking. Moreover, the whole device is very similar to Papin's own Digester, the

---

[18]James P Muirhead, *The Life of James Watt with Selections from His Correspondence* (New York: D Appleton & Co, 1859). http://books.google.com/books?vid=0sKexydQgpLl_a&id=0NSFx3lV-vAC&printsec=titlepage&dq=The+life+of+James+Watt

Fig. 4.5 Papin's 1690
steam apparatus (Adapted
by Thurston in *A History of
the Growth of the Steam-
Engine* (New York:
D. Appleton and Company,
1878), 50. 1878 edition:
http://www.history.
rochester.edu/steam/
thurston/1878/)

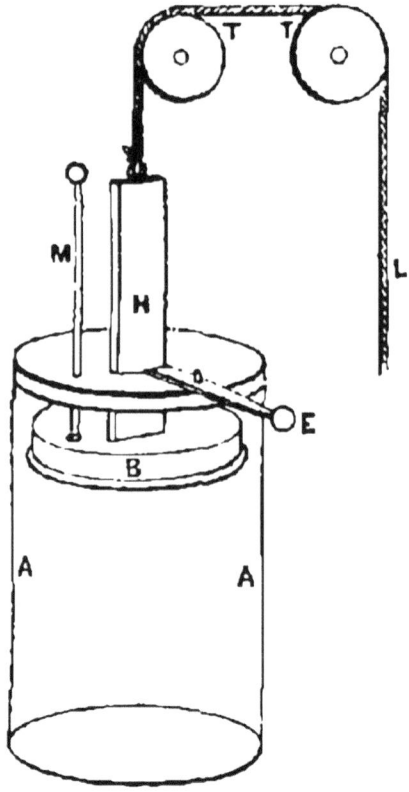

origin of the modern pressure cooker. The principle of expelling air to create a
vacuum was suggested by Huygens, so the idea of expelling it with steam might
have been judged, in patent examiner's terms, "prior art", that is, "obvious to
someone with a good knowledge and experience of the subject".[19]

The novelty of Papin's device lies in the combination of all of these into a
functional system that could operate in a cycle to do useful work. In Papin's cycle,
energy is supplied when the applied heat generates steam for the upward stroke and
atmospheric pressure to do useful work on the down stroke. Because the work is
done by atmospheric pressure not the steam, it is more accurately described as an
atmospheric engine rather than a steam engine.

Papin's principle was subsequently developed into a practical engine by Thomas
Newcomen. Because of their geographic locations, educational and social positions,
Newcomen, an ironmonger, probably never meet Papin nor read his paper (which

---

[19]US Department of Commerce, Manual of Patent Examining Procedure, (Alexandria, VA: US
Department of Commerce, 2001), http://www.uspto.gov/web/offices/pac/mpep/index.htm. web
page. 706.02.

Fig. 4.6 Newcomen's
atmospheric engine
(Thurston, *A History of the
Growth of the Steam-
Engine*, 59)

was in Latin), but Newcomen did know Thomas Savery who in turn knew Papin and had himself experimented with steam and patented a steam driven pump.

The engine Newcomen invented is shown in Fig. 4.6. Instead of heating and cooling the same quantity of water inside the cylinder as Papin did, Newcomen introduced a separate boiler to produce steam continuously, the steam being condensed to create a vacuum by spraying cold water directly into the cylinder. As with Papin's apparatus, the power to drive Newcomen's pump came from atmospheric pressure on the top of the piston. Like Papin's, Newcomen's is an atmospheric engine, the steam serving to create the required vacuum.

With these basic descriptions of Huygens, Papin's and Newcomen's inventions we can turn to the question of whether each is novel or merely a rearrangement of its predecessors. Compared with previous experiments on vacuums created by pumps, the novelty of Huygens device is his introduction of an energy source to produce a vacuum directly in the cylinder. Viewed as a functional system, it consists of the cylinder and piston, non-return valves pulley and weight, and gunpowder as energy source, all related by the motion produced.

Papin's innovation on Huygens's functional system was to replace a gunpowder explosion with steam and condensation. Papin's engine would not have been very efficient because of the time taken to heat and cool the fixed amount of water in the cylinder, but unlike Huygens sketch design, was a viable system consisting of the physical components in Fig. 4.5 and the accompanying cycle. Papin's functional system is shown diagrammatically in Fig. 4.7.

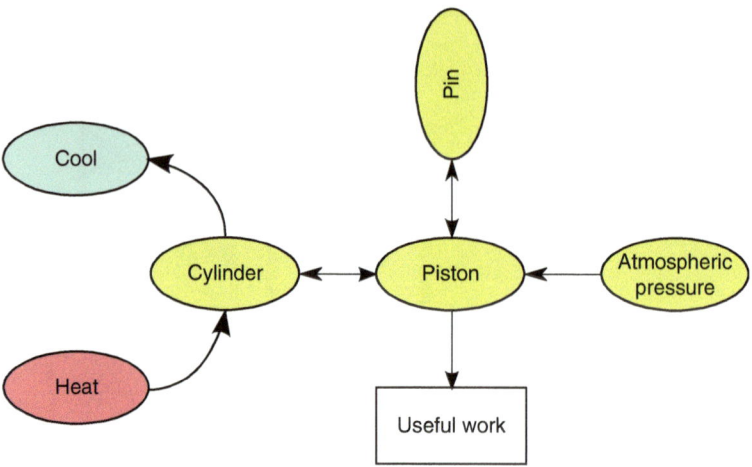

Fig. 4.7 Papin's atmospheric engine as a functional system

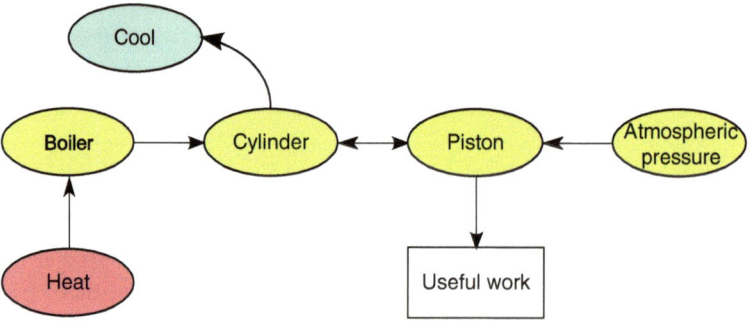

Fig. 4.8 Newcomen's atmospheric engine as a functional system

On superficial examination, Newcomen appears to use the same functional system as Papin, with steam raising the piston and condensation creating a partial vacuum causing atmospheric pressure to do useful work. If we limit Newcomen's functional system to this, his engine is not novel. However, when we look at the detail of Newcomen's functional system (Fig. 4.8) we can see that Newcomen expanded Papin's functional system and added new components. Disregarding the beam and pump in Fig. 4.6 which are the means of utilising the useful work analogous to Papin's pulleys and rope, we can see that Newcomen added a crucial component, the external boiler. Newcomen's functional system comprises the cylinder, piston, steam, air pressure, *plus* the external boiler and water injection. Replacing the fixed volume of water in Papin's cylinder with a boiler and using water injection to condense the steam increased the cycle speed significantly compared with Papin's machine. In functional system terms, the boiler and water condensation are not components present in Papin's functional system. Newcomen created a different and novel functional system.

If we seek novelty by looking for differences in components or operating principle, Huygens, Papin and Newcomen are essentially the same, differing primarily in their respective degrees of effectiveness. Newcomen may have used the same principle as Papin (atmospheric pressure to produce useful work with a cylinder and piston), but he did it by creating a novel functional system. Functional systems provide a way of distinguishing between the three and identifying the novelty in each through the differences in their systems.

Chapter 5
Innovation Must Fail

I didn't have much faith that [my first tinfoil Phonograph] would work, expecting that I might possibly hear a word or so that would give hope of a future for the idea. Kruesi, when he had nearly finished it, asked what it was for. I told him I was going to record talking, and then have the machine talk back. He thought it absurd. However, it was finished, the foil was put on; I then shouted "Mary had a little lamb", etc. I adjusted the reproducer, and the machine reproduced it perfectly. I was never so taken aback in my life. Everybody was astonished. I was always afraid of things that worked the first time. Long experience proved that there were great drawbacks found generally before they could be got commercial; but here was something there was no doubt of.[1]

5.1 Innovation

In this passage Thomas Edison captures the essence of this chapter: we should expect innovations to fail, because with innovations, failure is the norm. Indeed, failure is so fundamental to innovation that we should be suspicious if an innovation does not fail. Edison was "never so taken aback in my life" when his first attempt at a working Phonograph did not fail. Elsewhere he is quoted as saying "out of a hundred experiments he does not expect more than one to be successful, and as to that one he is always suspicious until frequent repetition has verified the original results".[2] There are fundamental reasons why Edison expected innovations to fail. They are the same reasons why we also should expect innovations to fail and be suspicious - or even, as Edison's says, afraid - when they do not.

In principle, an innovation is simply something new but the word has acquired connotations implying much more than that. Microsoft founder Bill Gates claims

---

[1]Edison quoted in Dyer and Martin, *Edison, His Life and Inventions*. 208.
[2]Ibid., 2: 612.

© Springer Nature Switzerland AG 2019
I. Wills, *Thomas Edison: Success and Innovation through Failure*, Studies in History and Philosophy of Science 52, https://doi.org/10.1007/978-3-030-29940-8_5

that "For centuries people assumed that economic growth resulted from the interplay between capital and labour. Today we know that these elements are outweighed by a single critical factor: innovation. Innovation is the source of US economic leadership and the foundation for our competitiveness in the global economy."[3] For Gates and for many others, to be innovative is not just a matter of choice, but an imperative, a question of economic survival. Yet Edison's reaction to the unexpected success of the Phonograph suggests that successful innovation does not come easily, that it requires much more than novel ideas.

5.2   Inventions, Functions and Means

Inventions can be thought of as a combination of the functions they perform and the means by which they achieve those functions. For inventors, questions of function are essentially *what* questions, that is, what to invent. Questions of means are essentially *how* questions; how to achieve those functions.

Edison began work on the Phonograph (Chap. 7) by identifying a function to be met, initially (and mistakenly) intending to invent a means of recording and reproducing telephone messages. He then identified a phenomenon (inscription of sound on waxed paper) that could be used as a means to achieve that function. It was the function, recording and reproducing sound, that Edison directed Edison's efforts and remained constant while the means for meeting it, the artefact he called the Phonograph, developed. Because Edison was pursuing a function, he was willing to abandon specific aspects of the artefact during its development in order to achieve the functions. He began by using wax as the recording medium, abandoned wax for tinfoil, patented the Phonograph based on tinfoil, then a decade later, returned to wax. Throughout this process, his objective, the function, remained constant: to record and reproduce sound.

This primacy of function in invention is reflected in the wording of Edison's patents. After some legal formalities, Edison's first Phonograph patent begins by setting out the Phonograph's functions:

> The object of this invention is to record in permanent characters the human voice and other sounds, from which characters such sounds may be reproduced and rendered audible again at a future time.[4]

It was this combination of functions, recording sound and later reproducing it that defined the Phonograph and made it a revolutionary invention. Others had recorded sound before Edison but the Phonograph was the first to reproduce it. As a

---

[3]Bill Gates. "How to Keep America Competitive." *The Washington Post*, 25 February 2007 2007. http://www.washingtonpost.com/wp-dyn/content/article/2007/02/23/AR2007022301697.html.

[4]Edison. Phonograph or Speaking Machine.

revolutionary invention, its functions are more significant than the details of how they were to be achieved. The patent fixed the artefact's functions and once fixed, Edison had commercial control of the exploitation that followed.

In a similar way, Alexander Graham Bell gained the controlling patent for the telephone, not because he had an effective device when he applied for the first patent (he did not) but because he patented the fundamental function of the telephone, transmission of sounds over a wire using a varying current. Bell's patent described one means for achieving this but whether something is patentable depends on whether it is novel, not whether it is effective at achieving the claimed novel functions. Once the commercial objective of a controlling patent for the function has been secured, the invention's means of achieving those functions can be improved. Other inventors may solve some problems of means, as Edison did with the carbon microphone, but Bell's control of the telephone through his patent meant that only Bell's companies could exploit it, so eventually the rights to Edison's carbon microphone patent ended in Bell's hands.

The next paragraph of Edison's first Phonograph patent describes the means of achieving its functions:

> The invention consists in arranging a plate, diaphragm, or other flexible body capable of being vibrated by the human voice or other sounds, in conjunction with a material capable of registering the movements of such vibrating body by embossing or indenting or altering such material in such a manner that such register marks will be sufficient to cause a second vibrating plate or body to be set in motion by them, and thus reproduce the motions of the first vibrating body.[5]

In terms of the invention's functions, the key phrases in this paragraph are, "capable of registering the movements of such vibrating body" and, "reproduce the motions of the first vibrating body". That is, to record and reproduce the vibrations of the human voice or other sounds. The remainder of the paragraph deals with the other aspect of invention, the means by which these functions are achieved. Edison's patent expresses the principle that an artefact, in this case an invention, is a combination of its functions and the means by which they are achieved.

Kroes and Meijers describe an artefact as having a "dual nature".[6,7] That is an artefact exists both as its functions and the physical manifestation of those functions. Kroes and Meijers confine themselves to what they refer to as "technical artefacts", a restriction not necessary for the present discussion, since, although the examples we are using are mostly technical, the conclusions drawn are applicable to non-technical artefacts. Further, the approach used here, unlike that of Kroes and Meijers, does not tie an artefact's functions to its physical manifestation because, as Chap. 3 argued,

---

[5]Ibid.

[6]Peter Kroes and Anthonie Meijers, "The Dual Nature of Technical Artifacts - Presentation of a New Research Programme," *Techné: Research in Philosophy and Technology* 6, no. 2 (2002).

[7]Mitcham also questions some other of the "dual nature" model of artefacts in Carl Mitcham, "Do Artifacts Have Dual Natures? Two Points of Commentary on the Delft Project," ibid.

the functions and success criteria for an artefact are not limited to those determined by its creator.

5.3 Novelty, Functions and Means

Functional success in inventing requires successful matching of means to functions. The form of words used in of introduction to Edison's Phonograph patent identifying means and functions is found in most patents, not just Edison's. The marrying of function and means in patents implies that we can have a novelty through novel functions, through novel means, or through a combination of novel functions and novel means. Edison's patent for a novel means of preserving fruit used a high vacuum pump which he had developed in connection with electric lighting.[8] Having devised a means for producing a high vacuum, Edison added the novel function (preserving food) to create a new patentable invention. That is, he attached a novel function to an existing means.

Edison also did the reverse. Other inventors had produced light from electricity before Edison, so his objective in inventing in electric lighting involved finding novel means to satisfy this existing function, achieved as we saw earlier, by adding a novel success criterion (high filament resistance) and the means for achieving it. Finally, occasionally Edison created inventions with both novel functions and novel means. The Phonograph was novel both in its functions (recording and reproducing sound) and the means of achieving these functions (inscribing on tinfoil or other soft medium).

Sometimes Edison started work on an invention with a function in mind and sought a means to achieve it. In other cases he started with a known means and sought a new function for it. On 17 July 1877, Edison identified a function to be fulfilled, the need to record telephone messages, and speculated on a potential solution.[9] Edison did not, on that day, have the means to achieve it. He had a function without means. By the following day, after his first sound recording experiments, he had both function and a means for achieving it, inscribing sound on waxed tape.

In other instances Edison started with a means then sought uses for it. Most commonly this approach started when Edison noticed a novel phenomenon then looked for functions it could be used to achieve. The astronomical instrument, the Tasimeter, that Edison took to Wyoming to measure the temperature of the sun's corona began in this way. Edison had noticed a novel phenomenon in a failed carbon microphone experiment, and sought functions that could be achieved with it, employing in this astronomical instrument and a few months later as a means for

---

[8]Thomas A Edison. Preserving Fruit. US Patent 248,431, filed 14 December 1880, and issued 18 October, 1881.

[9]TAEB 3:969.

regulating electric lamp filament temperature. With the Tasimeter he went from function (microphone to convert sounds into variable resistance), to phenomenon (Tasimeter principle), to new functions (astronomical instrument and filament temperature regulation). A similar sequence is identifiable with the Phonograph: a function (the need to record telephone messages for later transcription) led to Edison noticing the Phonograph phenomenon (inscription of sound in soft medium), which in turn led to a new function, recording sound for entertainment.

Such sequences like these occur many times in Edison's inventive process. His magnetic ore extraction venture began with the observation that some kinds of sand were magnetic (a novel phenomenon). This prompted him to try to invent a means for extracting iron from ore magnetically (a function), which in turn led to various inventions relating to processing of minerals (means). When magnetic ore extraction turned into a financial disaster, he adapted these mineral processing inventions and the knowledge gained from the venture to the manufacture of Portland cement (an old product but a new function for Edison's inventions). The manufacture of Portland cement, in its turn, led Edison into concrete manufacture and its use in prefabricated houses and larger structures (a new function). The chance observation in 1880 that some sand was magnetic led in 1923, though this sequence of steps, to the construction of Yankee Stadium from Edison's concrete.

Edison's habit of pursuing such linkages contributed to his prodigious inventive output. In addition to almost 1100 patents, Edison proposed many thousands of ideas for inventions that never got to the patentable stage. Viewed in hindsight, many clearly had no potential, while others with more potential never eventuated, usually because Edison either lost interest or turned his energy to other inventions.

5.4 The Phonograph as Novelty

Chapter 4 demonstrated the use of functional systems to identify novelty in inventions. This chapter adds another approach to identifying novelty: analysis of functions and means. Like Edison's Phonograph, Scott de Martinville's Phonautograph recorded sounds on a cylinder using a horn to amplify the sound energy. In terms of their components, the Phonograph and Phonautograph were very similar. What distinguishes the Phonograph is its ability to reproduce the recorded sound. The Phonautograph was an interesting scientific instrument that enabled sound to be visualised, but the Phonograph did much more. It was not just interesting, it was astonishing because it not only recorded sound but made those sounds audible as well as visible.

The ability to reproduce sounds made the Phonograph a revolutionary invention but most other inventions do not exhibit such radical novelty. Edison was not the first person to create an incandescent lamp coming after several dozen other inventors who produced lamps in the preceding decades, many using the same basic arrangement as Edison; carbon filament in a vacuum.

The novelty of the lamp Edison invented lay in the new functions he added to this frequently used configuration. Edison started work on electric lighting on 27 August 1878 and within two weeks (10 September 1878), was referring to "Electric Light Subdivision".[10] Edison made the "subdivision" of electric light his objective, not just electric lighting. For Edison, subdivision meant both reducing the intensity inherent in existing electric arc lighting so it was suitable for use in small rooms and subdivision of power from a central station to serve many buildings. Subdivision was a function, and Edison was seeking a means to achieve it. He soon concluded that achieving the second kind of subdivision required a lamp with a high resistance filament. Edison's innovation was the addition of high filament resistance as a success criterion to the function of producing light from electricity by incandescence. It was the critical step that gave Edison the controlling patent in electric lighting.

5.5 Success Criteria and Novelty

Subdivision of light was the function Edison sought to achieve, and a high resistance filament was the means by which he achieved it. Inventors state the claimed functions of their invention in their patent applications but an invention's functions are not exclusively those described by the inventor nor are they inherent in the invented artefact. The artefact embodies means (for example, high filament resistance) but the artefact's functions are also determined by those who interact with it in some way, no matter how remotely. The inventor's claimed functions are just one set of possible functions.

In the first electric light patent in which Edison mentioned high resistance and subdivision of light he wrote, "The object of this invention is to produce electric lamps giving light by incandescence which lamps shall have high resistance, so as to allow of the practical subdivision of the electric light".[11] While these were Edison's stated functions, for William Vanderbilt, who financed Edison's development of electric lighting, the function of this lamp was to create profit.[12] For the consumer, its function was to provide light. Since Edison's incandescent lamp had different functions for each of these, each judged it against different success criteria. Among Edison's success criteria was, "the practical subdivision of the electric light". Vanderbilt's success criteria included the rate of return on his investment. Consumers, whose success criteria included the quality and cost of electric lighting,

---

[10]TAEB 4:1426.

[11]Edison. *Electric Lamp*.

[12]William Vanderbilt (1821–1885) American Financier. Eldest son of Cornelius Vanderbilt, William took over his father's financial empire on his death in 1877 and was an early investor in Edison's electric lighting project which he saw as a threat to the Vanderbilt family's existing gaslight investments.

were probably not aware of Edison's or Vanderbilt's success criteria. Despite these differences, Edison's, Vanderbilt's and consumers' success criteria are interrelated by Edison's innovation of high filament resistance.

We saw in Chap. 3 that what is judged as success or failure in an artefact is not fixed but depends on the choice of success criteria and that the choice of success criteria can vary between people and may vary over time. Judgments of success and failure are further complicated by the fact that success criteria are often not expressed neatly in either/or terms but in relative or quantitative terms. When Edison began work on the telephone, he observed of one of his first instruments that, "you could tell someone was talking" but could not understand what they were saying.[13] That is, the instrument achieved the function of transmitting sound electrically but not that of transmitting intelligible speech. The carbon microphone that Edison eventually patented was far more effective. After hearing Edison's voice transmitted over a distance of 210 km, one observer commented that the quality was so good that, "I recognised your voice instantly".[14]

As with the incandescent lamp, Edison's microphone was not the first to be invented. What distinguished Edison's device was that, as well as transmitting intelligible speech, it also satisfied other success criteria that the Bell and Reis telephones did not. Unlike theirs, Edison's was compact, convenient to use, operated on existing telegraph lines and, because it was powered by the telegraph system, could transmit speech over hundreds of kilometres.

5.6 Novelty Lies in Identifying and Meeting New Success Criteria

Every artefact has a double, its success framework, used to judge whether the artefact successfully achieves all its functions. One artefact may have many success frameworks that may vary over time and between people who interact with it. Since inventions can be viewed as a combination of functions and means, success requires successfully creating both the means for meeting the required function and an appropriate success framework for judging its success. Because the success framework is itself an artefact and novelty can be achieved by identifying new success criteria, invention involves creating another artefact in the form of a new success framework. Consequently, novelty in inventions can also be the result of creating another novel artefact, the success framework.

If the functions of an artefact are well-established, its success framework will also be well-established so it is possible to achieve novelty by adding a new success criterion to an established success framework. From this perspective we can say that the novelty of Edison's invention of the carbon microphone and incandescent lamp

---

[13]TAEB 3:767n1.
[14]TAEB 4:1194.

lay in adding new success criterion to an existing success framework and devising the means to meet it. Edison's incandescent lamp not only produced light from electricity like those earlier inventors but his did it in a way that made it possible to supply electric lighting to many buildings from a central generating station.

Edison's addition of this new success criterion effectively made possible electric lighting as a utility. For this reason, it should not be his supposed invention of the light bulb that Edison should be remembered but the invention of electric lighting as a system.

Similarly, the novelty of Edison's carbon microphone lay in the quality of the sound transmitted and in its compatibility with the system in which it operated, a function missing from Bell's telephone. Concentrating on only some functions met by an invention (like producing light from electricity) can be misleading if we are trying to identify novelty. Examining success criteria that they meet offers an alternative means of distinguishing one invention from another and of identifying novelty.

5.7   No Functions, No Invention, No Means, No Invention

While identifying a new function led Edison to the invention of the Phonograph, failing to recognise potential functions could mean that he missed potential inventions. In 1875, Edison failed to convince many people that Etheric force was a new force of nature (Chap. 8). His understanding of Etheric force phenomena was shaped by his experience in electricity, so he visualised it as being transmitted by conduction like electricity, not realising he was working with wireless phenomena and so missed the opportunity to pioneer wireless telegraphy.

Etheric force was a notable exception because one of Edison's strengths was the ability to identify novel functions. Edison may have started working on the Phonograph with the mistaken belief that it would be used to record telephone messages but he soon modified this to the more valuable function of using sound recordings for entertainment. The outcome of each, missing wireless telegraphy and fame as the inventor of the Phonograph, was related to the way in which Edison matched a function to the means of achieving it. In the case of the Phonograph, Edison's match was successful. With Etheric force, not successfully matching function to the means meant no inventions and no patents.

The Phonograph and Etheric force are both examples of Edison finding (or not finding) a new function in a means he already possessed but the reverse is also evident. In November 1877, Edison correctly identified a new function for the Phonograph when he proposed mass production of tinfoil recordings so that a family might have one Phonograph machine and many tinfoil recordings of music. The new function of mass producing recorded music was perceptive, but the means, tinfoil recordings, proved impractical. Edison had correctly identified a function but was unable to develop the means of achieving it until he devised a new recording medium, the wax cylinder.

5.8   Who Invented the. . .?

It is not uncommon to encounter statements such as "Newcomen did/did not invent the atmospheric engine", "Edison did/did not invent the light bulb" and "Bell did/did not invent the telephone". In considering such claims we need to ask what we mean when we say that a particular person invented a particular artefact.

No invention exists in isolation from other inventions. That is, all inventions for which we have historical records embody earlier inventions and/or are based on earlier inventions. A number of writers including Basalla, Petroski and Ziman have referred to this as an evolutionary process although evolution, when applied to technology, lacks an identified driving process analogous to natural selection.[15,16,17] The point that these writers emphasise is that the origins of new technologies lie in earlier technologies some of which may have used natural phenomena. Basalla, for example, describes the development of barbed wire from the use of the thorny Osage Orange as a hedge to confine cattle.[18] Edison's Phonograph may have contributed a novel function (sound reproduction), but he developed it from a number of his existing inventions including the acoustic telegraph and automatic telegraph. Even without these, it is obvious from Edison's first demonstration Phonograph's that it made use of far older inventions, notably the screw and the crank. Rather than expressing Edison's priority through the simplistic phrase "Edison invented the Phonograph", we might more accurately say, "Edison combined a number of existing inventions into the Phonograph, by adding a novel function, sound reproduction, to the functions already achieved by earlier inventions including Scott de Martinville's Phonautograph". Or, instead of saying, "Bell invented the telephone" it would be more accurate to say that, "Bell invented the telephone by adding voice modulated variable electrical current to the existing two state (on-off) electric telegraph technology".

An examination of patents shows that most are improvements to existing technologies. Edison's most significant contribution to the incandescent lamp was not a new kind of lamp or materials, but the high resistance filament. That contribution was only significant because it allowed the lamp to become part of a larger electric lighting system. Edison's contribution was the realisation that for such an electric lighting system to economically serve more than one building as gas lighting did, its lamps needed to operate at high resistance. Without the high resistance filament lamp, electric lighting, as a large technological system, would not have been commercially feasible. This is the basis for the statement that Edison invented electric lighting as a system. Similarly, the telephone acquired its significance because the parts that Edison and Bell invented fitted into a technological system.

---

[15] Basalla, *The Evolution of Technology*.

[16] Henry Petroski, *The Evolution of Useful Things*, 1st ed. (New York: Knopf, 1992).

[17] John Ziman, ed. *Technological Innovation as an Evolutionary Process* (Cambridge, Cambridge-shire: Cambridge University Press, 2000).

[18] Basalla, *The Evolution of Technology*, 52.

For the first 10 years of its existence, Edison sold very few Phonographs. The Phonograph's impact came with the development of mass produced Phonograph recordings, which in turn created the sound recording industry as a technological system.

However technically minor an innovation may be, its impact may be to make a previously non-viable technology viable. Thus high filament resistance, a minor addition to the lamp, made electric lighting viable as a utility. This suggests that inventors who add a small but crucial innovation may become identified as *the* inventor of the artefact, despite not being the first to produce it. For this reason in most cases it is more accurate to say "X invented a crucial improvement in Y" rather than "X invented Y". The significance of the improvement is not necessarily related to its physical effect on the existing invention. High and low filament resistance lamps appear to be identical but the significance of filament resistance lies in how the lamp interacts with other parts of the larger electrical supply system.

Such minor improvements do not necessarily mean minor problems to be solved. As Edison summed it up himself, "It is easy enough to invent wonderful things and set the newspapers talking, but the trouble comes when you try to perfect your inventions so as to give them commercial value".[19] We can understand Edison's expression "commercial value" as success against commercial success criteria. For Edison, a successful invention was not a "wonderful thing" with some technical promise but one that had been developed to the point where it satisfied commercial success criteria.

"X invented Y" statements are further clouded when novelty is the result of adding new success criteria. To say, "Edison added the success criterion of high filament resistance to existing incandescent lamp technology", does not have the same impact as, "Edison invented the incandescent lamp", despite being more accurate.

These apparently minor changes can give their inventor the controlling patent that enables them to dominate the new industry the patent creates. Despite the potential for commercial dominance, controlling patents usually signal only the beginning of a development process for the invention, not the culmination. Such a patent may be controlling patent commercially but an enabling patent technically. This can be seen in the illustration in Bell's controlling telephone patent (Fig. 5.1). It bears little resemblance to any commercial telephone but Bell's innovation was not in the particular device illustrated in the patent but his description of the concept of sound being transmitted by a "vibratory or undulatory current of electricity in contradistinction to a merely intermittent or pulsatory current".[20] Possession of this patent may have given Bell commercial control of the telephone but it also enabled the development of other patents such as Edison's carbon microphone. Without Bell's patent there may have been some interest in the carbon microphone but not the demand that made it valuable as part of the telephone system.

---

[19]quoted in Conot, *A Streak of Luck*, 284.
[20]Bell. Improvement in Telegraphy.

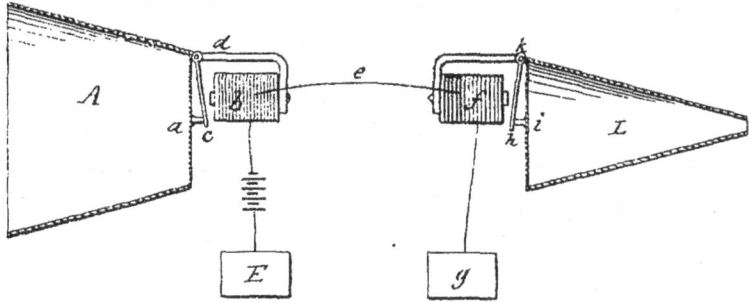

Fig. 5.1 Illustration of the telephone in Bell's controlling telephone patent. (Bell. Improvement in Telegraphy)

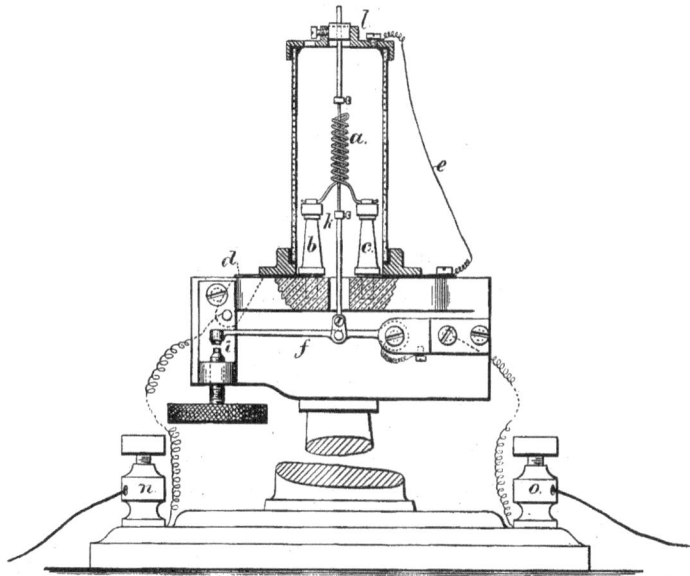

Fig. 5.2 The incandescent lamp illustrated in Edison's first electric lighting patent. (Thomas A Edison. Thermal Regulators for Electric Lights. US Patent 214,637, filed 18 November 1878, and issued 22 April 1879)

Similarly, the drawing of the incandescent lamp in Edison's first patent (Fig. 5.2) is barely recognisable as a lamp. It was not until his tenth electric lighting patent, the one in which he identified the need for a high resistance filament (Fig. 5.3), that he showed something we can recognise as the lamp's familiar form.

Even the concept of, "the inventor" does not fare well when closely examined. By law, the patents name one or more individuals as its inventors, they do not name organisations as inventors. Although Edison is named as sole inventor on most of his patents he did not invent alone. One of his associates, Francis Jehl, observed that "In

Fig. 5.3 Illustration from
Edison's controlling electric
lamp patent. (*Electric Lamp*)

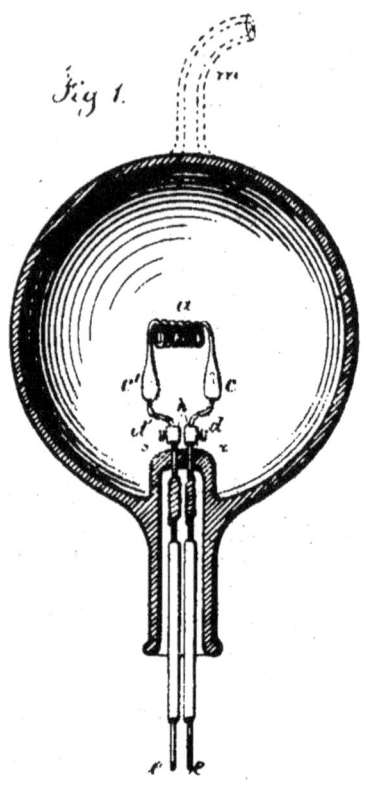

reality, Edison is a collective noun and meant the work of many men".[21] That is,
Edison invented as part of an invention-producing functional system. Even before
applying for his first patent, Edison was inventing with others including long-time
friend and fellow telegrapher, Ezra Gilliland.[22,23] Despite always working with
others, Edison rarely used suggestions that came from employees, in part because
to do so may have required including them as co-inventors. Edison may have been a
lone inventor in terms of patent law but he did not invent alone, men like Thomas
Batchelor and John Kruesi being essential to his success. The same was true of
Edison's contemporaries including Alexander Graham Bell, Elisha Gray and Elihu

---

[21]quoted in Conot, *A Streak of Luck*, 469.

[22]Ibid., 18.

[23]Ezra Gilliland (1846–1903) A telegrapher and inventor like Edison. Gilliland worked with Edison
on many early inventions, then later during Edison's Menlo Park period. Their friendship ended in
1889, as many of Edison's did, in acrimonious dispute. Gilliland moved to Edison's rivals Bell
telephone and then Gray's Western Union.

Thompson. It is also true of earlier periods. Galileo may have invented improvements to the telescope but, like Edison, part of its construction was undertaken by others. The statement, "X was the inventor of Y" is more accurately expressed as "X led a group of people who invented the enabling improvement to Y".

In summary, we can say that:

1. Inventions do not stand alone. They build on existing technologies and embody existing technologies.
2. Most inventions are minor changes to existing technologies.
3. Apparently revolutionary inventions tend to be revolutionary because of their impact on a technology (Hughes's reverse salients) rather than the extent of the changes to preceding inventions.
4. Many apparently revolutionary inventions are revolutionary because they make large technological systems feasible. Without the creation of such systems, these inventions might be little more than curiosities.
5. First attempts at inventions (and first patents) rarely represent the familiar form of the invention but can give their inventors legal control of the invention, inhibiting competitors and giving the inventor time to develop and patent more effective versions.
6. Often the innovation occurs not in the means by which functions are achieved but in identifying and meeting new success criteria for existing technologies.
7. Successful inventors rarely work alone. Inventions are rarely the creation of one person.

Given these factors, searches for the ultimate inventor of successful inventions are essentially futile. Neither Edison nor Swan was the first to produce light from electricity by incandescence. The lamp made by the Belgian, Jobard, in 1838 heads Friedel and Israel's list of incandescent lamp inventors but even he had antecedents including Davey and Singer.[24] Although Jobard, like Edison and Swan, used a carbon filament in a vacuum, he, like Davey and Singer, produced a device that had no commercial significance. When our notion of an incandescent lamp is of the device Edison developed to a commercial product as part of an electric lighting utility, it is misleading to say "Jobard invented the incandescent lamp" because although Edison's and Jobard's lamps operated on similar principles, Jobard's could not have functioned in an electric lighting utility. Indeed, the incandescent lamp we currently envisage is not the lamp illustrated in Edison's tenth electric lighting patent, or even the lamp in Fig. 5.3 that first which brought electric light to New York offices in September 1882.

Understanding the futility of searches for an ultimate inventor is important because it allows us to re-examine the development of inventions and so identify the innovations embodied in key inventors.

---

[24]Friedel, Israel, and Finn, *Edison's Electric Light: Biography of an Invention*, 115.

5.9   Invention as System Creation

Chapter 9 argues from the patterns in Edison's patents that he did not start with a grand vision of a large technological system and then invented components to build it. Instead he had a flexible approach that allowed him to move functional system boundaries frequently to produce new systems and to identify previously unrecognised combinations as new functional systems or to apply these new systems to new functions. New systems were opportunities for new patents. Patents, not systems, drove Edison.

The process of invention has two separate but related aspects. The first is the creation or identification of a function to be fulfilled. Since a function is the objective of a functional system, this implies a search for a functional system. Edison's initial jury-rigged Phonograph experiment described in Chap. 7 contained all the components of the Phonograph as a functional system: the diaphragm recording point, recording medium and movement, components that persisted in one form or another for the next 70 years until the advent of electronic Phonograph pickups in the late 1940s. The second aspect of inventing involves developing the functional system (or, more accurately, a series of functional systems) into a viable, patentable, device.

5.10   Why Do Some Innovations Not Fail?

Despite Edison's many notable successes as an innovator, in terms of patents, many more of his innovations failed than succeeded. Analysis of innovation from a functional system perspective shows that failure is inherent in innovation. The more complex the system an innovation represents, the greater the number of interactions within it and so the greater the opportunity for failure. From this, we could expect that failure to be a certainty yet there are many examples of innovations that have not failed. Indeed, innovation would hardly have the allure that it has if there was no hope of eventual success. So, why do some innovations not fail?

The answer to this question is relatively simple. Except for rare cases like the first Phonograph, innovations that do not fail, succeed because the adverse interactions that would have caused them to fail have been met and addressed, or have been anticipated and addressed. As with artefact creation generally, the key to success lies in anticipating and overcoming failure, that is identifying success criteria and success clues and devising ways of satisfying them.

While this may appear straightforward, it raises the question as to how such adverse interactions are anticipated. In mature technologies, like building a new bridge or an aeroplane, they are anticipated because most of the relevant success criteria and success clues are well-established, sometimes codified, so potential adverse interactions are confined to the small part of a total system that has not been done before: the innovative part. Not only are the success criteria well-established in mature technologies, but the techniques for dealing with the kind of minor innovations encountered are also well-established, for example mathematical and physical modelling, prototyping and commissioning.

This implies, as we have seen many times, that much accumulated knowledge is built on the analysis of past failures. In mature technologies this may represent decades or even centuries of accumulated failure examples. The task facing a major innovation, that is a major departure from established technology, is to acquire this kind of knowledge. Since this knowledge will also largely come from the analysis of failures, it implies an extensive development process in which seeking and overcoming failure are significant. In effect, at the end of such a development process, the innovation, to its innovators, has ceased to be an innovation and has acquired the attributes of a mature technology. Because of this, the amount of development required to produce a failure-free innovation is related to the degree of innovation: the greater the innovation, the greater the development and failure seeking required.

This has two consequences. Firstly, radical innovation without extensive development will probably fail. It is not that successful innovations are impossible but rather that if success is to be achieved, it will come after many failures. If the innovation has not failed yet, it probably will.

Secondly, innovation that is claimed to be radical, but produced without either extensive development or without failure, most probably succeeds because it is actually a mature technology, for which the knowledge based on past failures is well-established. That is, the innovation is either minor or perhaps even cosmetic.

5.11 Innovation and Risk

Radical innovation, something that is radically new, inevitably means a lack of knowledge of what might be problematic about many of the interactions the innovation involves. The effect of this lack of knowledge is that innovators cannot address the problematic interactions and consequently we should expect the innovation to fail. Except for extremely rare cases like Edison's first Phonograph, if something is claimed to be innovative and doesn't fail, it is probably because there is little about it that is new, that is, there is little innovative about it.

Individuals, organisations and societies that value continuous innovation also need to expect and value frequent failure, and to be sceptical of claimed innovations that do not fail.

We also saw that the more complex the innovation, the greater the number of interactions and the greater the number of opportunities for it to fail. In contrast, simplicity increases the probability of success because it means fewer interactions. It is a concept inherent in many of Edison's most successful inventions. Compare the complexity of the electric lamp in Fig. 5.2, his first patent, with the simplicity of the lamp in Fig. 5.3, his tenth patent. It is this value of simplicity that is embodied in the KIS principle: "Keep It Simple ". This is such a pervasive idea that we understand it almost intuitively. Cartoons of fanciful inventions like those by Rube Goldberg, Heath Robinson and Gerard Hoffnung amuse in part because we intuitively know that the ridiculously complex inventions they depict will fail to do what they are meant to.

If innovation is so fraught with failure, it is worth considering whether there is an alternative. This question is usually only addressed in innovation rhetoric in the negative, often by the implication, that a lack of innovation means stagnation. Yet the alternative is identifiable from the essential meaning of the word innovation. Innovation is a relative term. Something is innovative if it includes something new, yet no innovation can be totally new and must be built on existing technologies. In that light, innovation becomes a question of degree, not an absolute. We can increase the chances of success and reduce the risk of failure by seeking simplicity and by making as much use as possible of mature technologies for which success criteria are well-established. This approach to innovation might be described as craft or craftsmanship. That is, excelling at what has been successful previously.

Chapter 6
Catastrophic Failure

6.1 Recognising Catastrophic Failure

Previous chapters have dealt with failure using examples drawn mainly from Edison's laboratory notebooks. Even when these failures had a significant effect on the inventions he was developing, none could be described as a catastrophic failure. In contrast, catastrophic failures are easily recognised because the artefact is either destroyed or severely damaged, often at considerable economic and human cost. It takes no specialist knowledge to recognise that a collapsed bridge, a train crash or a ship sinking is a failure. Because of the cost of catastrophic failures, most literature relating to the failure of artefacts deals with catastrophic cases yet the overwhelming majority of failures are not catastrophic or even dramatic. A pen that stops writing is just as much a failure as a collapsed bridge. Both the pen and the bridge are failures because they do not meet a success criterion, not because of the consequences of not meeting it. What distinguishes a catastrophic failure are the consequences of failure.

This chapter looks at catastrophic failures, applying the principles developed in previous chapters to them and highlighting some special aspects of catastrophic failures.

6.2 The Space Shuttle Challenger

The loss of the Space Shuttle Challenger in 1986 illustrates the way in which the same failure can be non-catastrophic (i.e. have non-catastrophic consequences) many times and then become a catastrophic failure.

The US Space Shuttles were launched using a combination of rockets: a liquid fuelled main rocket and two solid fuel booster rockets. So that the boosters could to be transported from their factory in Utah to the launch site in Florida, they were built

© Springer Nature Switzerland AG 2019   97
I. Wills, *Thomas Edison: Success and Innovation through Failure*, Studies in History and Philosophy of Science 52, https://doi.org/10.1007/978-3-030-29940-8_6

in sections, the joints between the sections (known as field joints) being sealed with O-rings made from a rubber-like material. In solid fuel rockets, the interior of the rocket and hence the seals to be subjected to the substantial pressure from the burning fuel. The Presidential Commission (known as the Rogers Commission) that investigated the explosion of the Challenger concluded that during its launch, the O-rings on the booster rockets failed to seal allowing burning gases to leak out. The leaking gases eroded the O-ring seals, increasing leakage until ultimately the escaping hot gases caused main rocket to explode.[1] While O-ring seal failure had catastrophic results for Challenger, hot gas leakage (also called blow-by), had been observed in two thirds of booster rockets recovered after earlier Shuttle launches.[2] Since blow-by and seal erosion indicate a seal that has not sealed, it is reasonable to describe these events as seal failure, yet until it caused the loss of Challenger, this seal failure was not catastrophic failure. Acceptance of seal failure on earlier Shuttle missions is an example of revision of success criteria, the failures being referred to as "anomalies". In effect, those responsible changed the success criterion of zero seal leakage to an unstated one in which leakage was permissible provided it did not have undefined serious consequences.

Success is success because it is not failure. As obvious as this is, it has a consequence that is not often acknowledged: in order to succeed we need to keep failure in mind. After a second Space Shuttle exploded in 2003 (the Columbia), the report into that disaster observed that, "Organizations that deal with high risk operations must always have a healthy fear of failure".[3] It is evident that before the 1986 Space Shuttle Challenger explosion, there were aspects where NASA did not have this healthy fear of failure; otherwise it would not have treated O-ring seal blow-by as non-failure. In making this observation, the Columbia Investigation Board was incorrect in singling out high risk operations. In doing so it implied that there are situations where failure can be tolerated. Any situation where we want something to succeed warrants a healthy fear of failure, or more accurately, a constant alertness to failure.

Of the 135 missions in NASA's Space Shuttle program, two ended in catastrophic failure with the loss of 14 lives, a failure rate of roughly one in seventy. It is a failure rate that would not be tolerated in other situations. There would, for example, be no commercial airline industry if commercial aircraft had anything like this failure rate. In fact, the reverse is true. Commercial air travel has such a low failure rate that catastrophic failures, when they occur, are prominent news stories and require detailed investigation. Commercial air travel is not alone in this. We find nothing remarkable about a building or bridge that has stood without failing for centuries.

---

[1]United States. Presidential Commission on the Space Shuttle Challenger Accident., *Report to the President*, 5 vols. (Washington, DC: The Commission, 1986), 40. https://spaceflight.nasa.gov/outreach/SignificantIncidents/assets/rogers_commission_report.pdf

[2]Ibid., 66.

[3]Columbia Accident Investigation Board, *Columbia Accident Investigation Board Report*, vol. 1 (Washington, DC: National Aeronautics and Space Administration and the Government Printing Office, 2003), 190. URL 7 Sept 2007: www.hss.energy.gov/deprep/archive/oversight/caib_report_volume1.pdf

One reason for this apparently high success rate is that creators of successful artefacts are, for the most part, constantly aware of the potential of failure and constantly seeking to avoid it. They are also helped by the fact that, for the most part, they are not dealing with innovations in complex systems.

6.3   Catastrophic Failure in Complex Systems

By their nature, the consequences catastrophic failures create intense interest and have produced a substantial literature both within engineering and more generally. Some catastrophic failures have national and international consequences like the toxic gas leak at the Union Carbide Bhopal plant in 1984, the Chernobyl nuclear plant explosion in 1986 and the loss of the Space Shuttles Challenger and Columbia in 1986 and 2003. Each of these represents the failure in a complex system involving mature technologies where unanticipated interactions led to catastrophic results. Such failures, because of the magnitude of their consequences, become exemplars for similar technologies and so have led to extensive analysis.

Perrow agues from analysis of catastrophic failures in large and complex systems including Bhopal, Chernobyl, and the Space Shuttle Challenger, that since such systems involve many interactions, inevitably at times such interactions will lead to failure.[4] If several failures occur simultaneously they can lead to further adverse interactions, the consequences of which were not, and could not, be anticipated. Because of this, Perrow argues, "Operator error" cannot be a valid explanation in these circumstances because the unpredictability of such interactions means that no operator could deal with them in the time available to do so.

Weick, Sutcliffe and Obstfeld offer an opposing view, pointing to what they describe as High Reliability Organisations (HROs).[5] These include nuclear power plants and air traffic control in which, they claim, a state of "collective mindfulness" prevents catastrophic failures. In effect, they believe that in HROs, collective mindfulness means that any departure from norms is treated as a failure requiring corrective action.

Both views are concerned with complex technologies but, as discussed in Chap. 4, even a relatively small system can have a large number of interactions. For this reason, the consequences and strategies each describes are also applicable to much smaller systems. Both are also concerned with mature technologies for which many past failures have provided knowledge including success criteria and success clues that can be used to avoid future failure. However, in innovative technologies that depart from established technology there will inevitably be less relevant

---

[4]Charles Perrow, *Normal Accidents: Living with High-Risk Technologies*, 2nd ed. (Princeton: Princeton University Press, 1999).

[5]Karl E Weick, Kathleen M Sutcliffe, and David Obstfeld, "Organizing for High Reliability: Processes of Collective Mindfulness," *Research in Organizational Behavior* 21 (1999).

knowledge from past failures to use to avoid future failures. This failure knowledge deficit increases as innovation increases.

With new technologies the failure knowledge deficit can be reduced, as Edison did, by testing and probing for failure but there are limits to this. A major problem confronting an innovator is that such tests can only be devised for situations where it is believed that failure might occur and that this knowledge is largely informed by past failures. We cannot test for failure in situations in which failure has not yet been imagined.

Further, such large scale testing can be complex and costly so frequently it is not done on the whole system and instead testing is performed in a modular fashion on subsystems. This subsystem approach has the advantage that it assists in identifying the causes of failures that might be difficult to locate if the whole artefact is tested but a common reason for testing only subsystems is that it is cheaper than repeated testing of the whole. Pinkus et al. argue that a contributing factor in the explosion of the Space Shuttle Challenger in 1986 was that, under budgetary pressure, only subsystems rather than whole assemblies of the Space Shuttle and its rockets were tested.[6] The problem with not testing the whole, combined, system is that the whole system will have many more interactions than simply the sum of the interactions in the subsystems. As a result, unexpected interactions between subsystems (and between components and subsystems of subsystems) may not be anticipated and addressed.

The risk of failure caused by interactions between apparently unrelated appears to be a lesson not learned from the 1986 loss of the Space Shuttle Challenger because the 2003 failure of the Space Shuttle Columbia was the result of an unanticipated adverse interaction between apparently unrelated subsystems. In 2003, it was between the insulation subsystem on the main rocket and an apparently remote subsystem, the Shuttle's heat protective tiles. Failure of the insulation as the rocket was jettisoned on launch caused pieces to strike the heat protective tiles. During launch and orbiting, the tiles have no function so the damage to them was not yet a failure. However, on re-entry days later, the damage allowed the intense heat of re-entry to bypass the protective heat shield causing the Shuttle to disintegrate. As with the booster rocket O-ring seal leakage, failure of the main rocket insulation had been noted on previous launches but the success criterion had been revised, in part because the failure was not yet catastrophic.

An alternative to testing the whole artefact is to test a theoretical, often mathematical, model of it. While such models may allow many more scenarios to be tested, they present problems analogous to the testing physical artefacts. Theoretical models also rely on assumed success criteria and expectations of potential sources of failure. This is compounded by the inevitable simplifications necessary to enable the models to be built and manipulated.

---

[6]Rosa Lynn B Pinkus et al., *Engineering Ethics: Balancing Cost, Schedule, and Risk - Lessons Learned from the Space Shuttle* (Cambridge, Cambridgeshire; New York: Cambridge University Press, 1997).

6.4   Eclipsing the Weakest Link

A common metaphor applied to catastrophic failures is that of a chain breaking at its weakest link. Examination of many catastrophic failures shows that it is a misleading, potentially dangerous, metaphor. After the 1986 loss of the Space Shuttle Challenger, President Regan convened a commission to investigate its causes.[7] Among the commissioners was Richard Feynman, a distinguished theoretical physicist, Nobel Laureate and popular science author and lecturer. Like Thomas Edison, Feynman knew the value of a striking demonstration to persuade. In a widely publicised televised press conference, Feynman demonstrated how a small O-ring that was flexible at room temperature became hard at the freezing point of water, a temperature similar to the atmospheric temperature when the Challenger was launched.

It was widely reported that Feynman had discovered *the* cause of the explosion. According to the weakest link metaphor, the cause of the disaster was the properties of the O-ring material. The metaphor implies that if only NASA (or more correctly the Morton Thiokol, the booster rocket manufacturer) had used a better material, the disaster would not have happened. Unfortunately, the reasons for the disaster were far more complex. While Feynman was a distinguished physicist, he was a novice at both aeronautical engineering and catastrophic failure investigation. Feynman may have succumbed to the Kruger-Dunning effect in which people with low levels of knowledge or skill overestimate their knowledge or skill while conversely those with high levels of skill or knowledge underestimate it.[8] (Edison may also have suffered from the Kruger-Dunning effect when he embarked on his ill-fated magnetic iron ore extraction venture, discussed in Chap. 10.)

Just because Feynman claimed to have identified one cause of the failure, it did not mean that it was the only clause. Eventually, when Space Shuttle launches resumed, not only had the O-ring material been replaced, but the whole design of the field joints had been revised changed. If the O-ring material were the only cause and the weakest link metaphor valid, there would have no need to change the joint design. The field joints causal were not the only problem. The Rogers Commission identified many other causes of the explosion including post-launch reviews of previous missions that had noted previous seal failures but not acted; systemic problems on the day of the launch; and the decision to launch on an unusually cold day. To this can be added factors including the existence of the field joints themselves, a product of the rockets being built across the country in Utah, and not near the launch site in Florida and the hazards of scaling up a previously unproblematic rocket design for the Shuttle.

---

[7]United States. Presidential Commission on the Space Shuttle Challenger Accident., *Report to the President*.

[8]Justin Kruger and David Dunning, "Unskilled and Unaware of It: How Difficulties in Recognizing One's Own Incompetence Lead to Inflated Self-Assessments.," *Journal of Personality and Social Psychology* 77, no. 6 (1999). (also referred to as the Dunning-Kruger effect)

What happened on 28 January 1986 was that all these factors – materials, design, management and low atmospheric temperature – fell into alignment with the Challenger launch. All had existed in the past; none had previously caused a catastrophic failure. Rather than the weakest link, a better metaphor is that of an eclipse. The earth and moon constantly orbit but occasionally their orbits come together in a way that causes a solar or lunar eclipse. Nothing has changed; eclipses are a product of the sun-earth-moon system. Similarly, the catastrophic failure of the Challenger was a product of the Space Shuttle program as a system in which the atmospheric temperature interacted with the O-ring material; the O-ring interacted with the field joint design, and so on. The Challenger failure was the product of an eclipse of these interactions, not the failure of a single weakest link.

6.5   In Complex Systems, Innovation Will Probably Fail

Chapter 4 showed that even simple systems can have many internal interactions and that the number of interactions increases rapidly as the size of the system increases. Many interactions will have no consequences while others will have consequences that cause failures. In mature systems most of these adverse interactions will have been encountered previously and the ways of preventing failure established. When innovations involve complex systems, many of the adverse interactions will be novel and so not able to be anticipated. As a consequence innovations involving complex system have a high probability of failure. We do not have the means or knowledge to anticipate all potential ways in which the components might interact or ways in which artefact as a whole might fail. Without this we cannot develop appropriate success criteria and hence cannot test for them and strengthen the artefact before it is placed in service. Testing is hard and testing innovative technology is especially hard.

Chapter 5 argued that we should expect, and even want, innovations to fail. In complex systems this is even more so and in innovative complex systems more so again. When the innovative complex systems have the potential for catastrophic failure, the risk of failure is so high we may have to consider what amount of catastrophic failure will be tolerated. At the very least such innovations require a high level of Weick et al's "collective mindfulness" particularly intense vigilance to detect and address non-catastrophic failures rather than revising success criteria so failure ceases to be failure.

Part II
Edison, Science and Invention

Chapter 7
Inventive Success: The Phonograph

7.1 The Phonograph Introduces Itself

On 7 December 1877, Thomas Edison and his associate, Charles Batchelor, visited the editor of *Scientific American* bringing with them "a little affair of a few pieces of metal, set up roughly on an iron stand about a foot square". To the editor's astonishment, the "little affair", Edison's first demonstration Phonograph, "inquired as to our health, asked how we liked the Phonograph, informed us that *it* was very well, and bid us a cordial good night" (Fig. 7.1).[1]

Uproar followed as more and more people crowded into the editor's office to hear the Phonograph perform. A week later, *Scientific American* published an account of the Phonograph, summing up the impression that it left with the comment, "No matter how familiar a person may be with modern machinery and its wonderful performances, or how clear in his mind the principle underlying this strange device may be, it is impossible to listen to the mechanical speech without his experiencing the idea that his senses are deceiving him".[2]

The Phonograph transformed Edison's public image. Before the Phonograph, he was known as a minor, if successful, inventor prone to making extravagant claims, like his Etheric force theory. After the Phonograph, Edison was still known his extravagant claims but the Phonograph showed that he could back them with astonishing inventions. The impact of the Phonograph was profound. In 1879, Edison's first biographer claimed, "No invention in the world's history has engendered more curiosity than the Phonograph".[3] The Phonograph became Edison's favourite invention and the one he worked with longest. Over his lifetime, he was

---

[1] Scientific American. "The Talking Phonograph." *Scientific American*, 22 December 1877, 384–85.
[2] Ibid. Also in TAEB 3:1150.
[3] J B McClure, *Edison and His Inventions* (Chicago: Rhodes & McClure, 1879), 75.

© Springer Nature Switzerland AG 2019
I. Wills, *Thomas Edison: Success and Innovation through Failure*, Studies in History and Philosophy of Science 52, https://doi.org/10.1007/978-3-030-29940-8_7

Fig. 7.1 Drawing of the
Phonograph from the
Scientific American article
announcing Edison's
invention in 1877 (Scientific
American. "The Talking
Phonograph." *Scientific
American*, 22 December
1877, 384–85)

awarded 178 Phonograph related patents, applying for his first three weeks after
visiting *Scientific American* in 1877 and the last in 1926.[4,5]

In the months that followed the *Scientific American* demonstration, Edison and
another of his associates, Edward Johnson, demonstrated the Phonograph across the
United States where it was received as near miraculous, to the extent that some critics
claimed that the sounds they heard from it were produced by ventriloquism.[6,7]

At the invitation of Joseph Henry, one of America's most distinguished electrical
scientists, Edison visited Washington in April 1878 to address the National Academy of Sciences.[8] Later the same day he gave US President Hayes a private

---

[4]Edison. Phonograph or Speaking Machine.

[5]Method of Producing Sound-Record Tablets. US Patent 1,690,159, filed 5 October 1926, and
issued 6 November 1928.

[6]Conot, *A Streak of Luck*, 109–10.

[7]Edward Hibberd Johnson (1846–1917) Inventor and Edison associate. As manager of the Automatic Telegraph Company, Johnson hired 24-year-old Edison to work on automatic telegraphic
instruments. Describing himself as an electrician, Johnson was an inventor in his own right and
demonstrated Edison's telephone (including the "musical telephone") in public concerts. Johnson
provided feedback and suggestions to Edison on improvements to the telephone and was subsequently to become a partner in Edison's electric lighting companies. Johnson, when vice president
of the Edison Electric Light Company in 1882, is credited as being the first person to decorate a
Christmas tree with coloured electric lights.

[8]Joseph Henry (1797–1878) American scientist who made fundamental discoveries in electricity
and magnetism. His prominence as a scientist enabled him to promote innovations in electrical
technology, including those of Edison and Bell. Henry later became the first director of the
Smithsonian Institution.

demonstration and on the following day, demonstrated the Phonograph in the Capitol before "a large company of ladies and Senators".[9,10]

Edison's address to the National Academy of Sciences was widely reported, the *Washington Post* describing it as a "Genius before science" in "a scene … that will live in history".[11] On the other side of the Atlantic, the London *Evening Star* described Edison as a "modern magician".[12]

The Phonograph led to frequent visits to Edison's Menlo Park laboratory by journalists keen to report on his latest miracles. It was after one of these that the *New York Graphic*'s William Croffut gave Edison the title, The Wizard of Menlo Park.[13] Other journalists dubbed him "the Napoleon of science" and "the inventor of the Age" but it was Croffut's title that stuck, Edison remaining the Wizard of Menlo Park long after he abandoned that laboratory.[14,15,16]

Not only did the Phonograph improve Edison's public image, it increased his prestige in the scientific community and led to invitations to address other scientific meetings and to join a scientific expedition to Wyoming in July 1878 to observe a solar eclipse.[17]

The Wyoming expedition was not a scientific success for Edison but it led to him discussing his work with other researchers. One of these was Henry Draper, an astro-chemist who had discovered oxygen in the solar spectrum and from this deduced that oxygen was present in the Sun.[18] During their discussions, Draper related how in 1847, his father, John Draper had constructed an incandescent electric lamp with a platinum filament. Like so many others, this lamp failed because the filament overheated, Draper's father concluding, "An ingenious artist would have very little difficulty in making a self-acting regulator, in which the filament should be maintained".[19] Edison seized the challenge and the day after his return from

---

[9]Neil Baldwin, *Edison: Inventing the Century* (Chicago: University of Chicago Press, 2001), 97–98.

[10]Evening Star. "The Phonograph at the Capitol." *Evening Star*, 19 April 1878.

[11]Washington Post, "Genius before Science."

[12]TAED MBSB1:171.

[13]William Croffut. "The Wizard of Menlo Park." *New York Graphic*, 10 April 1878.

[14]TAED MBSB1:117.

[15]New York Sun. "The Inventor of the Age." *New York Sun*, 29 April 1878.

[16]William Croffut (1835–1915) American Journalist.

[17]This invitation came through the efforts of George Barker, professor of physics at University of Pennsylvania, who had declined to support Edison's Etheric force claims.

[18]Henry Draper (1837–1882) American physician, astronomer and son of John Draper (1811–1882) a noted physician, scientist and inventor who devised improvements to Louis Daguerre's photographic process.

[19]quoted in Conot, *A Streak of Luck*, 120.

Wyoming began working in earnest on it, filing his first electric lighting patent application, for an automatic temperature regulator, 3 months later.[20]

7.2   The Beginnings of the Phonograph

The Phonograph was an exceptional device. Not only was it the origin of the recording industry but it was one of those rare inventions without precedent. The *Scientific American* article that announced its arrival mentioned sound recording instruments invented by several others including Marey and Rosapelly, Scott and Barlow.[21] (The Scott mentioned by Scientific American was Édouard Léon Scott de Martinville, inventor of the Phonautograph.) While these earlier instruments recorded sound, Edison's was the first to reproduce it. This revolutionary step seized the attention of those who heard it. Revolutionary as it was, Edison's first Phonograph was so simple that even people with no technical training could understand how it worked.

After Edison patented the Phonograph, Alexander Graham Bell regretted that he had "let this invention slip through my fingers" for "I had stated again & again in my public lectures the fundamental principles of the Phonograph. In showing to an audience the tracings produced by the Phonautograph I had said the motions indicated by the curves could be produced mechanically".[22]

7.3   Edison's First Phonograph Patent

Edison's first Phonograph patent included the drawing in Fig. 7.2 with views of the Phonograph from the side (marked "*Fig. 1*") and from above (marked "*Fig. 2*"). The drawing shows a helically grooved cylinder wrapped in thin tinfoil, the recording assembly consisting of a mouthpiece, diaphragm and scribing point on the left and a separate reproduction assembly consisting of an adjustable point, diaphragm and earpiece on the right. A screw fixed to the cylinder rotates in a nut P fixed to the mounting base. As the cylinder turns, the screw causes the cylinder to move to longitudinally under the recording point to inscribe the recorded sound on the tinfoil or under the reproduction point to or reproduce it. Edison's design for the Phonograph had developed considerably from his first crude waxed paper tape device (Fig. 7.3) but the instrument shown in the patent drawing is still very simple and

---

[20]Edison. Thermal Regulators for Electric Lights.

[21]Scientific American, "The Talking Phonograph."

[22]TAEB 4:1260.

Fig. 7.2 Illustration from Edison's first Phonograph patent (Edison. Phonograph or Speaking Machine)

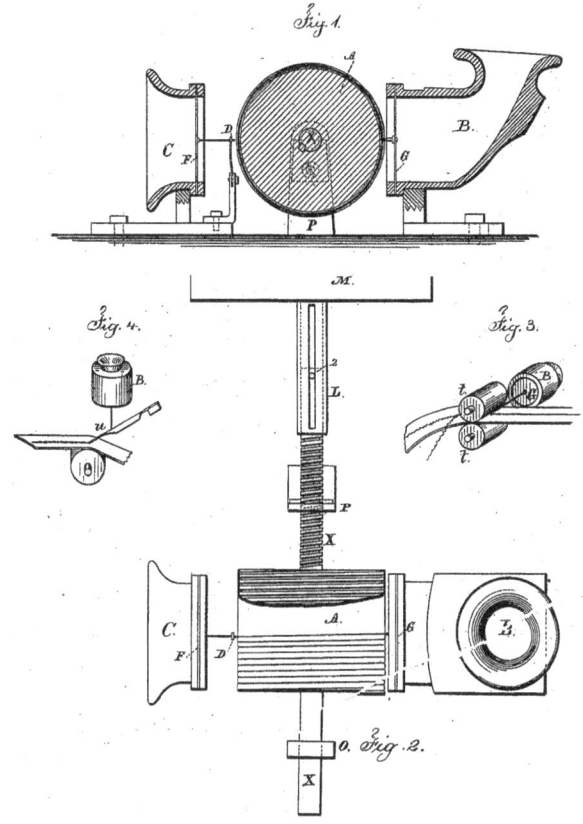

completely mechanical. Despite this simplicity, the basic principle described in this Phonograph patent, a sharp point recording sound as a groove in a soft material became the dominant recording technology for over a century.

7.4 The Invention of the Phonograph

In addition to its significance in the development of sound recording, several other factors make the Phonograph attractive for studying innovation more generally. Firstly, it is a rare kind of invention: one without precedent. In most situations, an inventor starting on a new project begins with a few examples to draw on, even if they not effective at achieving the intended functions. This was the case when Edison started on electric lighting, Draper's lamp being one example of an unsuccessful device. In the case of the Phonograph, Edison had no existing examples so

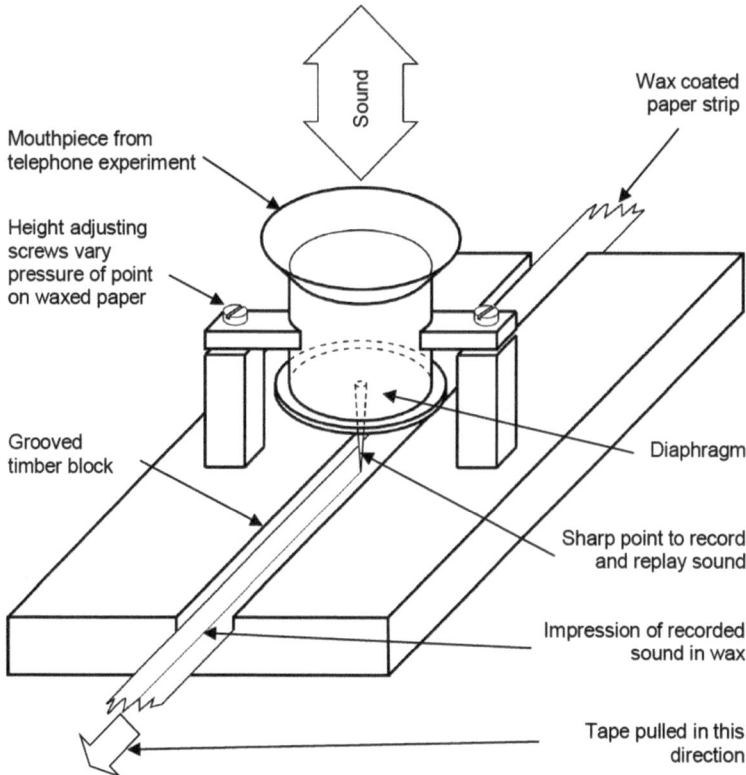

Fig. 7.3 Drawing of Edison's first Phonograph experiment based on Edison and Batchelor's descriptions

the development of the Phonograph was uninfluenced by pre-existing concepts or knowledge of other inventors' successes and failures.

Secondly, Edison's first Phonograph is attractive because of is its exceptional simplicity. It is far simpler than the carbon microphone, which in its final form was a very simple device. This simplicity facilitates study of the effects of variables on the functions it performs.

Finally, the Phonograph is attractive because of the historical records of its early development. These are contained in Edison's laboratory notebooks and in detailed legal testimony, he and Charles Batchelor gave in the 1890s describing its development. Although far from complete, these records give valuable insights into its early development.

This chapter offers two approaches to examining Edison's conception and early development of the Phonograph. The first draws on primary sources describing the development of the Phonograph, while the second is experimental, repeating Edison's first Phonograph experiments. The aim of the experimental approach is understand what led Edison to believe he could develop an effective sound recording and reproducing device.

7.5 Conceptual Origins of the Phonograph

Edison gave several, differing, accounts of the conceptual basis for the Phonograph. In one, he said the idea came to him from the sound made by an embossing telegraph repeater.[23,24,25,26,27,28] In another, it was the sound made by his Electromotograph machine.[29,30] A third version has it coming from the sound made by a disc recording.[31] Much more recently, Feaster argues for a very different origin: Edison's attempt to create a keyboard telephone that would have allowed users to "play" individual speech sounds over a telephone line rather than speaking them into a mouthpiece.[32]

While these are accounts of the conceptual basis for the Phonograph, it is important to distinguish the conceptual basis of an invention and the evidence for a viable means to achieve it. Although Alexander Graham Bell conceived the possibility of using the principle behind Scott de Leon's Phonautograph to reproduce sound, it was no more than a concept, something that might be possible, and a long way from even the earliest of Edison's sound recordings.

Conversely, having evidence of a means to achieve some new function is of little value without a corresponding conceptual base. In December 1875, Edison observed a "curious result" in his Etheric force apparatus, the Etheric force apparently transmitted over a distance of 2.5 m without wires (Chap. 8). Had Edison had

---

[23] *Edison, His Life and Inventions.* 206–07.

[24] *Menlo Park Reminiscences*, 1, 163.

[25] *Thomas Edison, Genius of Electricity*, Pioneers of Science and Discovery (London: Priory Press, 1974), 36.

[26] *Edison: The Man Who Made the Future* (London: Macdonald and Jane's, 1977), 74.

[27] *Edison: A Life of Invention*, 144.

[28] "Talks with Edison," *Harpers Monthly*, February 1890, 429.

[29] *Thomas Alva Edison: An American Myth* (Cambridge, Massachusetts: MIT Press, 1981), 12–13.

[30] Electro-Chemical Receiving Telephone. US Patent 132,455, filed 25 July 1879, and issued 31 August 1880.

[31] *TAEB 3*, 3, 695–97.

[32] Patrick Feaster, "Speech Acoustics and the Keyboard Telephone: Edison's Discovery of the Phonograph Principle," *ARSC Journal* 38, no. 1 (2007). The problem with this reconstruction is that Edison already held several patents relating to the automatic telegraph (stock ticker) which transmitted and printed characters directly. Such a keyboard telephone would have offered no advance over the automatic telegraph, since transmission might at best, have transmitted at a speed of around 120 characters per minute (say 20–30 words per minute). This is faster than a good Morse telegrapher and slower than the automatic telegraph, for which Edison claimed speeds of up to 1500 characters per minute from pre-recorded tape.

wireless telegraphy as a concept, he might have recognised the curious result as wireless transmission and developed it into wireless telegraphy. Instead, he thought of Etheric force only in terms of conduction and so missed this opportunity for a revolutionary invention.

7.6   The Phonograph Begins with a Misconception

Edison decided to invent a device to record and reproduce sound because he failed to recognise the telephone's potential for direct person-to-person voice communication. Edison, like most others working on the telephone in 1877, was an expert in telegraphy. Alexander Graham Bell was not, being primarily a teacher of the deaf. The result was that Edison and other telegraph experts saw the telephone as no more than an alternative method for transmitting telegrams, referring to it as the "speaking telegraph" whereas Bell recognised its direct person-to-person communication potential.[33]

As telegraph experts like Edison conceived the role of the telephone, a person wanting to send a message would present it in written form at a telegraph office. There, an operator would dictate it over the telephone to an operator in the receiving telegraph office who would down the message for delivery to the recipient. Since the sending operator could speak much faster than an even a good operator could tap out Morse code (around 100 words per minute spoken compared with 25 for Morse), the telephone had the potential to increase the speed of transmission substantially. It also eliminated the need for operators skilled in Morse code.[34]

On 17 July 1877, Edison noted a problem with this approach. As the sending operator could speak much faster than the receiving operator could write, the transmission speed would be limited to slower of the two. Edison did not consider using shorthand as a solution but it is notable that the term phonograph had been in use since the 1840s to refer the recording of speech by shorthand.[35] Instead, Edison's proposed solution was to record incoming voice messages mechanically. Operators at the receiving station would then play back the recorded messages, starting and stopping them to match their writing speed. Such a recording device could also be used when the receiving office was busy or even unattended. Edison already held patents for instruments that recorded conventional telegraph (Morse code) text messages for later playback so the sound recording device fitted that existing approach. In this notebook entry Edison sketched several possible approaches to recording sound including recording by indenting or perforating a strip of paper with

---

[33]Hounshell, "Elisha Gray and the Telephone: On the Disadvantages of Being an Expert."

[34]TAED NV12:8.

[35]Oxford English Dictionary, *"Phonograph, V."* (Oxford University Press). http://www.oed.com/

a needle, using a friction ink and a device that used a "Revolving plate [with] two telephone tubes".[36]

7.7   Edison's First Phonograph Experiments

The night after Edison made this speculation (18 July 1877), he conducted some experiments after which he wrote,

> Just tried experiment with a diaphragm having an embossing point & held against paraffin paper moving rapidly. The new speaking vibrations are indented nicely & there's no doubt that I shall be able to store up & reproduce automatically at any future time the human voice perfectly.[37]

This brief and incomplete notebook entry marks a very significant event: the first time a human heard recorded sound.

Edison liked to play with experiments in a manner akin to doodling. While waiting for questions to form in his mind or for the right experiment to present itself, Edison often passed the time by idly doing experiments in the general area of his concern.[38] Charles Batchelor's description of the events leading up to this experiment suggests that it was the result of this practice. In testimony in the 1890s, Batchelor said that Edison had been speaking into a telephone mouthpiece and feeling the vibrations of the diaphragm with his finger. (Since Edison was partially deaf, he frequently resorted to feeling when he his hearing was inadequate.) After doing this for a while, he turned to Batchelor and said "Batch, if we had a point on this we could make a record on some material which we could afterwards pull under the point, and it would give us the speech back".[39] Batchelor then fitted a knifepoint to the centre of the diaphragm and mounted the mouthpiece on a piece of wood so that a strip of paper coated in paraffin could be pulled under it. Batchelor continued, "On pulling the paper through a second time, we both of us recognized that we had recorded the speech".[40] In a slightly different version of the story published a few months after the event, Edison said that he had been feeling the vibrations with a point already attached to the diaphragm before he decided to try using it to record.[41] The result of this crude experiment convinced Edison that he had something worth developing.

---

[36]TAEB 3:969. For an analysis of these devices, see Feaster, "Speech Acoustics and the Keyboard Telephone: Edison's Discovery of the Phonograph Principle."

[37]TAEB 3:972

[38]Quoted in Hughes, "Edison's Method," 18.

[39]TAEB 3:972n4.

[40]TAEB 3:972n4.

[41]Brooklyn Daily Eagle. "Phonograph: A Machine That Talks and Sings." *Brooklyn Daily Eagle*, 26 February 1878.

Neither Edison nor Batchelor sketched the experimental apparatus at the time but Batchelor described it in 1896 using a sketch that has also not survived. In the absence of any original drawings, I have drawn one, based Edison and Batchelor's descriptions of the experiment and drawings in his notebooks of components Edison was working with at the time (Fig. 7.3).[42]

The first crude recording apparatus consisted of parts borrowed from existing inventions. It used a mouthpiece and diaphragm from Edison's current telephone experiments mounted on the piece of grooved wood from an old automatic telegraph. To make their first recording, Edison and Batchelor pulled wax coated paper tape along the groove and past the recording point while shouting into the mouthpiece. To play the recording, they pulled the tape under the recording point again causing the diaphragm to vibrate and produce sound. There are several accounts of the source of the wax coated paper tape. One describes it as recording paper from Edison's automatic telegraph while in another it is described as paper used for making electrical capacitors in which the wax served as a dielectric.

7.8   Edison Develops the Phonograph into a Patentable Invention

In 5 months, Edison transformed this crude device into the patentable invention that astonished *Scientific American* staff. Importantly, Edison not only transformed the device physically, he transformed his intentions for it. This second was crucial: the Phonograph moved from being a valuable but mundane adjunct to the telegraph office to being an exciting means of public and home entertainment.

Edison may have established a workable principle for sound recording on 18 July 1877, but developing it presented him with a set of related problems that appeared with each new invention. His first experiment convinced him that it was possible to record and reproduce sounds but this device was a long way from a commercially marketable invention. For that, he needed to find suitable materials and an arrangement to exploit them. A year earlier when he started to work on the telephone in earnest, Edison quickly determined that he could create a microphone that used a variable resistance, his initial device enabling him to "get a good many words Plain such as How do you do".[43] While this convinced him that he had the basic operating principle, it took over a year to find a suitable resistance material, carbon black, and almost 2 years to get a successful arrangement of components, the carbon microphone going into production in mid-1878. While the Phonograph demonstrated in December 1877, was a major advance on his first crude experiment, it was far from ready for sale to the public. Edison established companies to exploit his Phonograph

[42]TAED QP001:4 TAEB 3:972n4, Amos Jay Cummings. "A Marvellous Discovery." *New York Sun*, 22 February 1878.
[43]TAEB 3:759.

patent in 1878, but it was not until the end of the next decade that the Phonograph became commercially successful.

Compared to the carbon microphone, Edison's development of the Phonograph between July and December 1877 was remarkably rapid, particularly as he made only thirteen laboratory notebook entries. This suggests he spent very little time on the Phonograph, as he made hundreds of notebook entries on the telephone in the same period. Despite the shortage of notebook evidence, in 1896 Edison and Batchelor testified that they had recorded sound experimentally on many different materials during the second half of 1877.[44] The following account of Edison's Phonograph work draws primarily on the limited contemporary notebook entries supplemented by the testimony that Edison and Batchelor gave in 1896.

In his 1896 testimony, Batchelor described various ways in which they made sound impressions in the recording medium. Several times, he described the recording point as a knifepoint, noting that with soft materials like wax, the recording was made by cutting away the wax leaving fine shavings, but with harder materials such as tinfoil, the impression was made by embossing or indenting.[45] He added that in the early experiments using wax it was Edison's intention to cut the wax using a knifepoint or a point with an edge on it, rather than using a rounded point to displace the wax. (The patent infringement litigation in which Edison and Batchelor testified was concerned with the exact nature of the waxy material used, the kind of recording point and the nature of the recorded indentation.)[46]

There are no contemporary notebook records describing the paper tape used, but in his testimony, Batchelor said that they cut it from sheets of "condenser paper", that is, waxed paper for making electrical capacitors. This paper came in sheets about 18 by 36 inches (about 0.45 by 0.9 m) from which they cut strips to suit the width of a grooved block they already had.[47] In this testimony, Batchelor contradicted his earlier evidence that they had used automatic telegraph paper but "waxed paper" is consistent with Edison's 18 July notebook entry referring to "paraffin paper".[48,49]

The automatic telegraph paper Batchelor referred to in his earlier evidence came in large rolls and fitted the width of the grooved block they already had. As Edison had built a machine to produce waxed condenser paper, it is feasible that he could have used it to apply paraffin wax to automatic telegraph paper.

By the time they gave their testimony in 1896, Edison's and Batchelor's recollection of their experiments on 18 July 1878 may well have been clouded by the variety of materials they tested over the following few months. Given this, both

---

[44]TAEB 3:1013n1.

[45]TAED QP001:56 57.

[46]American Graphophone Company versus Edison Phonograph Works.

[47]TAED QP001:63.

[48]TAED QP001:64.

[49]TAEB 3:972. Edison described the paper as "bibulous" by which he seems to have meant that it was absorbent.

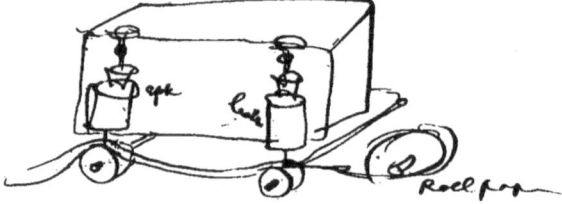

Fig. 7.4 Edison's sketch of 12 August 1877. The device uses paper tape and separate recording and reproduction heads labelled "spk" (speak) and "listen" (TAEB 3:1004). This device is based on Edison's automatic telegraph, a connection that possibly led Batchelor to later recall them using automatic telegraph paper rather than condenser paper

condenser paper and wax coated automatic telegraph paper are feasible for the 18 July experiments.

During July and August 1877, Edison was preoccupied with work on the carbon microphone so it was almost four weeks before he mentioned the Phonograph again in his notebook. On 12 August 1877, he sketched an arrangement in Fig. 7.4 using a roll of paper tape with separate devices for recording and replay.[50] His description of the mechanical drive indicates that some development had occurred, at least conceptually, as the manual tape transport of the original device is replaced by a mechanical paper feed. The notation however adds "any power to rotate" and the absence of experimental results suggest that this was a sketch for later construction rather than a device already built. In testimony, Batchelor said that they had adopted this as a solution to the problem of pulling the paper tape through at the same speed for recording and reproduction.

In their testimony both Edison and Batchelor referred to "a vast number of experiments" they carried out around this time, Edison adding that "I have always been experimenting with the Phonograph more or less since 1877; I would drop it for a while, and then take it up; ever since 1877".[51] While this is an empathic claim, there are few laboratory notebook entries to support it, the key phrase perhaps being "more or less". Edison made this statement in connection with litigation over his Phonograph patents so it is possible that he was trying to enhance his claim to intellectual priority by exaggerating the amount of work done. Rather than having "always been experimenting with the Phonograph", his notebooks indicate that several years separated some of the periods of development, particularly after Edison became involved in his electric lighting project in the second half of 1878.

Batchelor's testimony however corroborates Edison's claim that they did many more experiments in 1877 than were recorded in the laboratory notebooks. Given the later significance of the Phonograph, it is unfortunate that Edison and Batchelor made so few records of their early Phonograph experiments.

---

[50]This document also includes a dated sketch marked "Kruesi make this Aug 12/77". Kruesi did not make such a device until December.

[51]TAED QP001:11, 12.

On 17 August, Edison made several notes relating to the Phonograph among others about the telephone.[52] He proposed using cork for the sending and receiving diaphragms, listed several recording mediums and speculated about other methods of recording including indenting a ridge previously embossed on paper and recording on the thin edge paper. "Another idea Indent the paper in spiral grooves or on a long strip cover whole with tinfoil. The point of the diaphragm will then easily indent".[53] This was the first time Edison mentioned tinfoil and spiral grooves, key elements of the Phonograph he demonstrated to *Scientific American*. In this notebook entry, Edison was still thinking in terms of waxed paper recording medium but in his later testimony, he said that he also thought of covering a waxed surface with tinfoil because he found that cutting the wax produced small chips that stuck to the surface, interfering with reproduction.[54] In this entry Edison still referred to the Phonograph in connection with the "speaking telegraph" (telephone) and included a number of other uses of the Phonograph including as a repeater for telephone signals and one with a compressed air amplifier connected to the Phonograph, a concept he later patented as the Aerophone.[55] The number of ideas sketched and their lack of detail suggests they are ideas for possible later development and not experiments tried.

There is a further gap of several weeks before Edison's next sketches, dated 7 September 1877, in which he developed some of the ideas described on 17 August.[56] This entry shows recording on paper tape using the edge of the tape, friction of ink on tape, embossing "or knock down previous boss", and the use of a piece of thread to be displaced by the recording point then pressed into the waxed surface, the thread being removed to leave an impression in the waxed tape. This latter approach would have reduced the energy needed for recording as Edison found cutting the waxed tape needed "a powerful voice" and that they had to "holler" into the device.[57]

The difficulty of providing enough energy to record sound was a recurring problem. In testimony, Edison said that during this period they were trying to achieve reproduction using a "funnel", that is, a bell mouth of the kind used on musical instruments. In his first patent application, he described one solution to this problem, which was to replace the funnel with stethoscope earpieces permitting recording at much lower levels. Edison marketed several Phonographs using this approach (Fig. 7.5). Despite mentioning this in the patent application, in his

---

[52]TAEB 3:1013.

[53]TAEB 3:1013.

[54]TAED QP001:11.

[55]Thomas A Edison. Improvement in Speaking Machine. US Patent 201,760, filed 4 March 1878, and issued 26 March 1878.

[56]TAEB 3:1040.

[57]TAED QP001:12.

Fig. 7.5 An Edison
Phonograph using multiple
stethoscopes

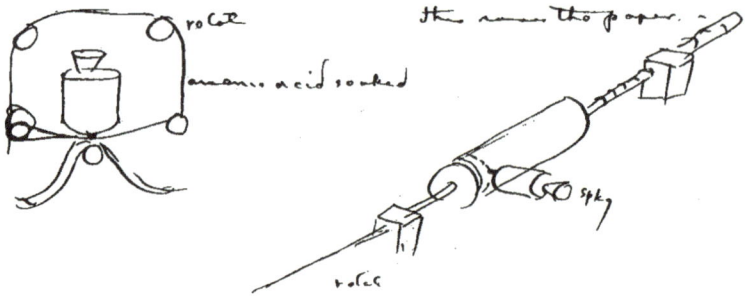

Fig. 7.6 Edison's first sketch of a cylinder Phonograph (TAEB 3:1062)

testimony Batchelor stated that during 1877 and 1878 they concentrated on Phonographs with funnels.[58]

A week and a half later, (18 September) Edison made several sketches of telephones and Phonographs, the Phonograph still using paper tape. Edison noted that they should use "mica & cork diaphragm in all".[59] Around this time, Edison was having some success using mica and cork diaphragms in the telephone microphone, nothing that "we made a lot of telephones this way and they all showed that it did the biz hunky Dori".[60]

Three days later (21 September), Edison sketched a cylindrical Phonograph for the first time (Fig. 7.6) among a large number of sketches of the telephone and Phonograph. In other sketches, he drew a system of recording that used arsenic acid to emboss paper, another that squeezed liquid from a tube to record sound and a third

---

[58]Edison. Phonograph or Speaking Machine.

[59]TAED TI2:269.

[60]TAEB 3:1062.

Fig. 7.7 Edison's
1 November Phonograph
sketch using wax coated
paper tape (TAEB 3:1099)

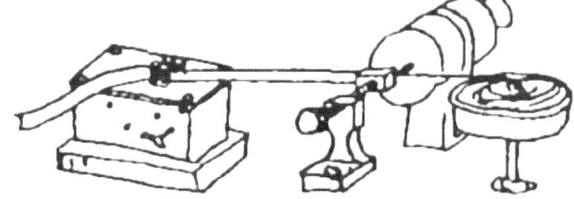

that used an electromagnetic microphone to generate an electrical current that in turn squeezed out fluid.

Yet another long gap followed these sketches, it being almost 6 weeks before Edison returned to the Phonograph on 1 November 1877. In this entry, he sketched the device in Fig. 7.7 that used paper tape "coated heavily with a compound of Beeswax & Paraffin, or other soft substance".[61] The left part of the sketch has what appears to be a clockwork mechanism wound by a key to draw paper tape from a roll on the right. The recording-reproduction device is in centre and resembles Edison's telephone designs of this period suggesting a close connection between the development of both.

In this entry, Edison also sketched what he described as a reed diaphragm, that is, a diaphragm fixed on one edge only, rather than all round, with a recording point (described as a "knife" in his 1896 testimony) on the opposite, free, edge.

Discussing the date of 1 November on this entry, Edison claimed in testimony that the dates of notebook entries were the dates on which they were signed adding, "The sketch could have been made a couple of months before that", implying that 1 November date on this sketch was not necessarily the date on which it was drawn.[62] While this might be true of other notebook entries, the sparse references to the Phonograph before 1 November make it unlikely he sketched this much earlier than this date. The number of significant notebook entries with November dates suggests that Edison did few experiments before November and that most of its critical development happened in that month.

Edison had mentioned using tinfoil in spiral grooves but later entries continued to refer to using wax-coated paper as the recording medium. However, on 5 November, he sketched a tinfoil Phonograph for the first time with this description: "On Phonograph I propose having a cylinder 10 threads or embossing grooves to the inch cylinder one foot long on this tinfoil of proper thickness arranged with cylinder is transmitter with embossing point running in these grooves or over them when cylinder rotated. I have tried various experiments with wax chalk etc to obtain an easy indenting surface but find that tinfoil over a groove is the easiest of all = this cylinder will indent about 200 spoken words & reproduce them from same

---

[61]TAEB 3:1099.
[62]TAED QP001:10.

cylinder".[63] It is a significant entry. Other than the length of the cylinder, this description is close to the first Phonograph built by John Kruesi at the beginning of December. There is no other evidence that he had built such a device, from wax, chalk or anything else at this stage, so it is likely that made the experiments mentioned using flat material.

The day after Edison made this entry (6 November) his associate, Edward Johnson, wrote to *Scientific American* describing the Phonograph. *Scientific American* published the letter in its next issue and was reproduced or commented on in other publications.[64] As with Edison's earlier and similarly startling announcement about Etheric force, newspaper reaction was mixed. Most applauded the news, the *New York Sun* noting that it would potentially enable the dead to speak to the living. In contrast, the *New York Times,* as it had done with Edison's Etheric force claims, lampooned the idea of the Phonograph. It prophesised that if the Phonograph became common, its owners would become as boring as those wine connoisseurs who drone on about their renowned vintages, Phonograph owners boring their guests with rare recordings of famous people.[65,66]

Johnson's letter was not the first public announcement before Edison had anything to demonstrate. In August, Johnson mentioned it in a public demonstration of the telephone in Philadelphia where it was reported in local newspaper.[67] Edison also referred to it in a British telephone patent application, probably drafted a few days after his 18 July experiments, in which he claimed that he could "make a record of the atmospheric sound waves".[68] Privately, Edison referred to sound reproduction in a number of letters. In one, to railway magnate Jay Gould dated 30 September, he claimed he had " ... nearly completed a machine which records the human voice on paper from which after the lapse of any time the same voice can be reproduced at any speed & with its fine inflection".[69] In another letter, to an associate, Benjamin Butler, he claimed he could "record the voice at 150 words per minute".[70]

On 10 November 1877, Edison made his first sketch of what was to be the familiar form of the Phonograph (Fig. 7.8).

---

[63]TAED TI2:342. If spoken at 80–100 words per minute (slow, loud speech being needed to produce a recording), this would have resulted in a recording time of 2–2.5 min, typical playing time of disc records up to the introduction of long playing records and still a common duration for pop music.

[64]TAEB 3:1102.

[65]New York Sun. "Echoes from Dead Voices." *New York Sun*, 6 November 1877. http://edison.rutgers.edu/singldoc.htm (TAED MBSB1:77).

[66]New York Times. "The Phonograph." *New York Times*, 5 November 1877, 4.

[67]TAEB 3:1103n4.

[68]TAEB 3:973.

[69]TAEB 3:1075.

[70]TAEB 3:1084.

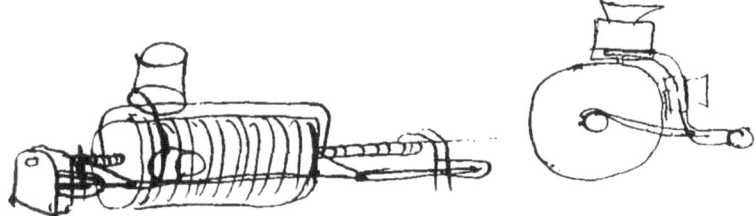

Fig. 7.8 Edison's Phonograph sketch of 10 November 1877. The notation reads "Phonograph tissue paper & tinfoil over it 10 threads to inch on Cylinder" (TAED TI2:367)

The proportions of the Phonograph in this sketch are similar to those of the instrument Edison was photographed with in Washington in April 1878 (Fig. 4.1) but the cylinder is thinner and longer than in the instrument he demonstrated to *Scientific American* in December (Fig. 7.1) and shown in his patent application (Fig. 7.2). In testimony, Batchelor said that Edison had the second built a few days after the first. Despite Kruesi's first Phonograph working much better than Edison expected, it appears it was not as he intended.[71]

A notebook entry dated 23 November, contains Edison's first proposal to market the Phonograph as a device for playing mass produced recordings. As he conceived it, a family might have a single Phonograph and 100 sheets of tinfoil containing a variety of recorded music. In the entry, Edison proposed mass producing recordings by embossing them onto tinfoil using a master recording in which the groove would be made larger (and hence louder) by engraving or by building up the metal using electroplating.

In this entry, Edison also proposed fitting miniature Phonographs inside "Dolls [that] speak sing cry & make various sounds also apply it to all kinds of Toys such as Dogs animals, fowls reptiles = human figures: to cause them to make various sounds".[72] In 1887, Edison created a company to manufacture and market dolls fitted with miniature Phonographs. The venture was unsuccessful as the Phonographs of this period were too delicate for such an application and the all-metal dolls too heavy for children to play with.

Although Edison left no notebook entries describing successful recordings at this time, he was confident enough of the principles of the Phonograph to order construction of a purpose built Phonograph to use instead of the makeshift devices he had been using up to that point. Edison drew a sketch of what he wanted built (Fig. 7.9) on 29 November from which John Kruesi made the first Phonograph. This instrument used a grooved cylinder, the pitch of the grooves matching the pitch of the driving screw (ten threads per inch). The top sketch in Fig. 7.9 shows the

---

[71]TAED QP001:50.
[72]TAEB 3:1119.

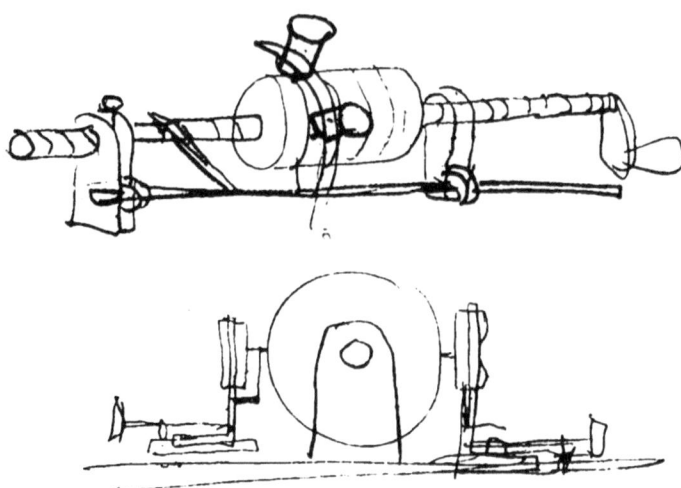

Fig. 7.9 Edison's sketch of 29 November 1877 from which John Kruesi built the first demonstration Phonograph (TAED NS77:3)

recording and reproduction heads on a common mounting while the bottom sketch shows them separate and on opposite sides of the cylinder, the arrangement Kruesi built.

While Kruesi was building his first Phonograph, Edison made a notebook entry on 3 December that included the comment "Have tried lot of experiments with different thickness of tinfoil. It's the best material yet for recording".[73] The same entry included sketches of three versions of Phonograph using a cylinder, tinfoil backed paper tape and, notably, a flat disc (Fig. 7.10), all three of which were included in Edison's first Phonograph patent.[74] Despite sketching a disc Phonograph, Edison preferred the cylinder Phonograph even after the introduction of disc recordings, Edison companies continuing to manufacture cylinder recordings until the 1920s.

---

[73]TAED NV17:21.

[74]Edison. Phonograph or Speaking Machine.

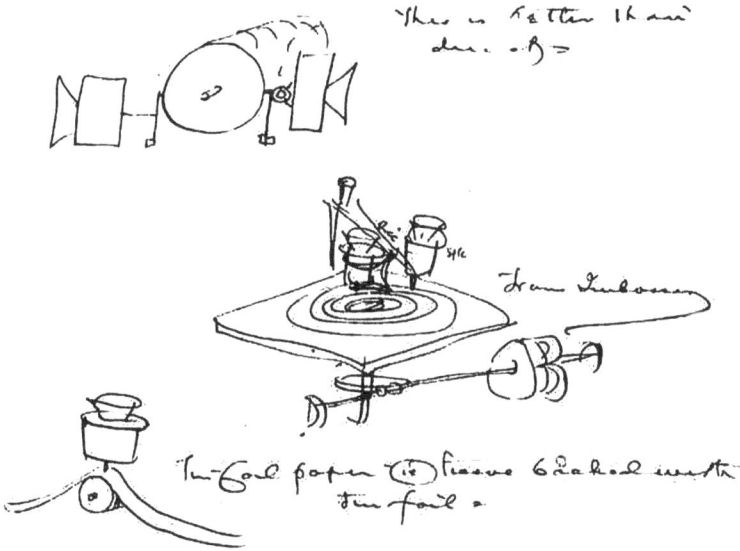

Fig. 7.10 Edison sketch of 2 December showing cylinder, disc and paper Phonographs (TAEB 3:1137, TAED NV17:21)

7.9   Kruesi's First Phonograph

John Kruesi built Edison's first tinfoil Phonograph between 1 and 6 December 1877. To make a recording, Edison wrapped tinfoil tightly around the cylinder and brought the point of the recording head into contact with it. The person making the recording then shouted into the mouthpiece while turning the handle to rotate the cylinder, advancing it under the recording head. As the tinfoil passed, the vibrations of the recording point indented the tinfoil over the groove. When the recording was finished, the record point was withdrawn and the handle turned in the opposite direction to return the cylinder to its starting point. To reproduce the recorded sound, the reproduction head was brought into contact with the tinfoil. When the handle was turned, the tinfoil moved past the point on the reproduction head and the indentations caused it to vibrate in response, producing sound from the diaphragm.

As simple as this device was, it was by no means simple to operate, requiring a combination of brute force and manual dexterity to achieve a satisfactory recording. Apart from the care needed when handling the delicate tinfoil and adjusting the recording and preproduction points, making a recording required both a loud voice to make an adequate impression and a steady hand to rotate the cylinder at constant speed. It appears from the *Scientific American* report of Edison's first demonstration, that he brought prepared tinfoil recordings with him, probably because of the awkwardness of recording.

Despite these problems, Kruesi's first purpose-built Phonograph was such a major advance over the apparatus Edison had been experimenting with he described his astonishment in the quotation that opened Chap. 5. It is significant and worth repeating:

> I didn't have much faith that [my first Phonograph] would work, expecting that I might possibly hear a word or so that would give hope of a future for the idea. Kruesi, when he had nearly finished it, asked what it was for. I told him I was going to record talking, and then have the machine talk back. He thought it absurd. However, it was finished, the foil was put on; I then shouted "Mary had a little lamb", etc. I adjusted the reproducer, and the machine reproduced it perfectly. I was never so taken aback in my life. Everybody was astonished. I was always afraid of things that worked the first time. Long experience proved that there were great drawbacks found generally before they could be got commercial; but here was something there was no doubt of.[75]

This is a telling statement for a man known for his great self-confidence, particularly given the significance the Phonograph was to have for Edison's reputation and career. It is also a significant revelation about the early development of the Phonograph raising two important questions: first, why was Edison so surprised that it worked and second, given that he seems to expect that it would not, why did it work? The first question is addressed in Chap. 5. Put simply, Edison – and we – should expect innovations to fail and be surprised when they do not.

One possible answer to the second question is that Edison was just lucky but as we saw in Chap. 4, the probability of getting a successful combination of components and their interactions even for such a simple device as this Phonograph is improbably low. Good luck is not a satisfactory explanation of why this first Phonograph worked. A much more credible answer is evident from Edison's development of other inventions. When he worked on the Phonograph, as elsewhere, he used failure as a tool and applied trial and error to develop systems of regularities that allowed him to predict what was likely to succeed. By the time he came to instructing John Kruesi to build this first Phonograph, Edison knew reasonably well the necessary details of recording medium, recording and replay point shape and so on. Given the novelty of the Phonograph there could be no relevant theory to help Edison, systems of regularities were all that anyone could have had at that time.

7.10   Repeating Edison's Early Recording Experiments

Edison's surprised reaction suggests that what he heard from Kruesi's instrument was far better than he was expecting, raising the question of what he had heard previously. His post-1877 accounts imply that he had been successfully recording and reproducing speech since July 1877 but if he had, why was he so surprised? Unlike the notebook entries for the carbon microphone with their many descriptions

---

[75] Quoted in Dyer and Martin, *Edison, His Life and Inventions*. 208.

of the sounds Edison heard, the entries relating to the Phonograph before December 1877 contain nothing. As this question had the potential to yield valuable insights into the origins of the Phonograph, I decided to try to answer it by repeating Edison's first experiments based on the available documents.[76]

Experimental approaches are well established in the history of science, Gooding Palmieri and Sibum for example, have applied the approach to the experiments of Faraday, Galileo and Joule respectively but experimental replication is much less common in the history of technology.[77,78,79] In what follows I describe my attempts to recreate Edison's early experiments. It was to be a challenging and time consuming project.

7.10.1 The Hand Phonograph Mark 1

My first step in repeating Edison's experiments was to build the hand held device he used to feel sound vibrations on 18 July 1877. (I will refer to it as the hand Phonograph.) Inspection of photographs and drawings of the telephone components Edison was working with around this time (for example, Fig. 7.11) indicated that it probably consisted of a parallel sided tube 100–120 mm long and 40–50 mm diameter, open at one end with a diaphragm at the other.

Using this information, I constructed a functionally similar device (Fig. 7.12), the tube being 120 mm long with a 43 mm outside diameter, flanged at one end to hold a diaphragm and with a sensing point (a 3 mm screw, ground to a point), fixed in the centre of the diaphragm.

7.10.2 The Hand Phonograph Mark 2

My first use of this suggested that the diaphragm was too stiff so, using trial and error, I tested various diaphragm materials and thicknesses including aluminium, copper, brass, steel and plastic in thicknesses varying from 0.1 to 0.5 mm. The best results achieved were with 0.1 mm copper sheet and with the retaining flange removed (Fig. 7.13). With this instrument, I could reliably feel the sound through the end of the sensing point by holding my finger against it. While I could detect the

---

[76]*American Graphophone Company Vs Edison Phonograph Works*,(1896).

[77]David Gooding, "Mapping Experiment as a Learning Process: How the First Electromagnetic Motor Was Invented.," *Science, Technology and Human Values* 15, no. 2 (1990).

[78]Otto Sibum, "Reworking the Mechanical Value of Heat: Instruments of Precision and Gestures of Accuracy in Early Victorian England," *Studies in History and Philosophy of Science* 26, no. 1 (1995).

[79]Paolo Palmieri, *Reenacting Galileo's Experiments: Rediscovering the Techniques of Seventeenth-Century Science* (Lewiston: Edwin Mellen Press, 2008).

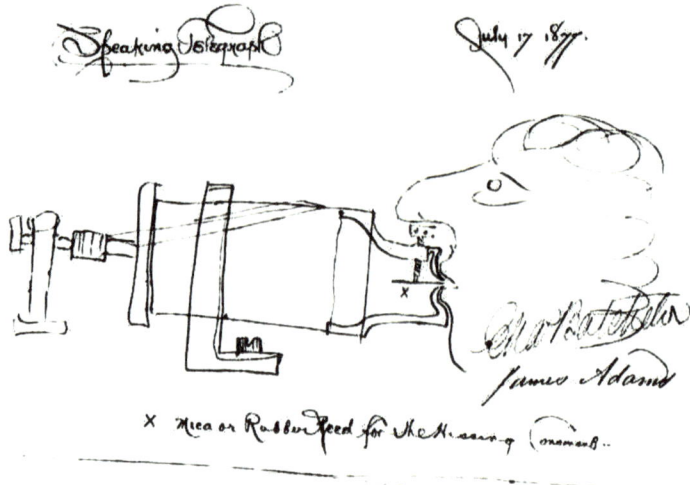

Fig. 7.11 Telephone transmitter Edison sketched on 17 July 1877. It has a reed added to the mouthpiece (TAED TI2:186)

Fig. 7.12 First hand Phonograph (Mark 1). It uses a plastic tube, fibreboard retaining flange and aluminium diaphragm

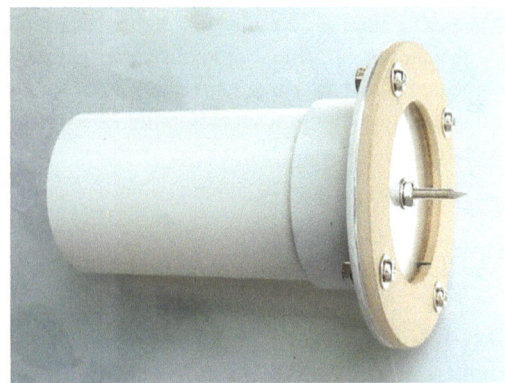

Fig. 7.13 Hand Phonograph Mark 2 using a copper diaphragm and no retaining flange

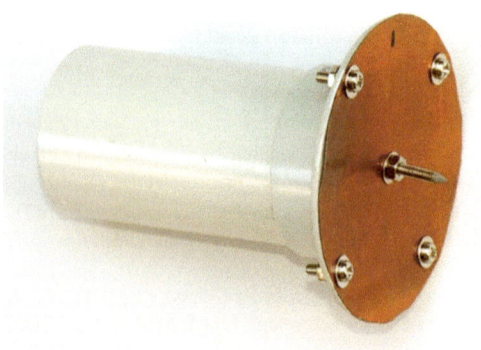

Fig. 7.14 The waxed tape Phonograph Mark 1

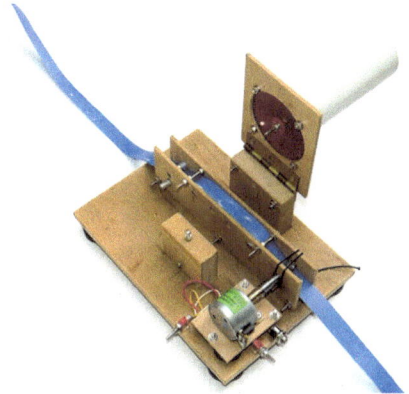

vibrations, the sensation was quite modest. In contrast, the reverse process (dragging my finger past the point), produced an impressively loud sound. After about 3 h constructing and testing the device in various configurations, I was satisfied that I had reproduced Edison's first experiment faithfully, achieving much the same results as he reported. Encouraged by this success, I moved to the next part of the replication, building the crude device described by Batchelor and illustrated in Fig. 7.3.

7.10.3 The Waxed Tape Phonograph Mark 1

At this point, I made what turned out to be an overly ambitious decision. Drawing on Edison's descriptions and the instrument sketched in Fig. 7.7, I decided to follow Edison and build something that resembled the tape transport in his automatic telegraph (stock ticker). The automatic telegraph printed stock prices on paper tape, advancing it a character at a time. To eliminate the effects of variations in tape speed, I adopted a mechanical drive like Edison but replaced his clockwork mechanism with an electric motor. Figure 7.14 shows the first device I built. I mounted the successful hand Phonograph on a hinged plate so it could be brought into contact with the recording tape, which moved through the instrument in a path similar to that of an automatic telegraph. The electric motor driving it is at the bottom of the photograph.

On testing this, I discovered what Edison probably found too: although the motor was quite capable of moving the tape with the recording point withdrawn, with the point in recording position the resistance created by cutting the wax was too great for the friction drive between the rollers and slippery waxed tape. The result was that either the tape did not move or, when it did, it moved erratically. To solve this, I removed the motor and resorted to Edison's original approach: puling the tape by hand.

7.10.4 Recording Medium

Having successfully tested the basic hand Phonograph and reproduced Edison's first crude device, the next task was to find a suitable recording medium. For this purpose, I sourced paper tape 15 mm wide, about the same width as Edison's automatic telegraph tape. In his first notebook entry, Edison refers the paper having a paraffin coating but elsewhere he refers to it as a paraffin-beeswax mixture.[80,81] Beeswax is a natural substance subject to the kinds of variation in properties typical of natural products, while paraffin is a generic term applied to a range of substances varying from brittle solids to liquids.[82] Consequently, the exact properties of the products used by Edison are unknown. To deal with this, I ran some trials with each wax in its pure form and in paraffinbeeswax mixtures of varying proportions, applying them to the tape by drawing it through the melted wax.

These trials showed that the form of paraffin I had was too brittle and stiff causing it to peel from the paper tape when bent. Pure beeswax on the other hand was flexible and did not peel but was too soft for successful recording. After some trial and error, I settled on a mixture consisting of one part beeswax to seven parts paraffin. A second variable was the thickness of wax for recording. In a notebook entry dated 1 November, Edison refers to paper tape "coated heavily with a compound of Beeswax & Paraffin, or other soft substance".[83] In his testimony, Batchelor said, "In the first experiments that I referred to, the paper was coated with paraffin wax, in which, to the best of my recollection, the thickness of the wax would be about twenty to twenty-five thousandths of an inch."[84] Assuming this thickness is per side, the total thickness would have been 1.02–1.27 mm. I came to a similar conclusion, using tape with the wax mixture applied in two coats and reduced to a uniform thickness of about 1.7 mm total using the scraping device in Fig. 7.15.

7.10.5 Recording Points

A major part of the testimony Edison and Batchelor gave in the 1890s related to the shape of the recording point, which they described at various times as "a point" and "a knife-edge". To determine which was the most suitable, I conducted a series of tests using a variety of shapes including conical and pyramidal, eventually settling

---

[80]TAEB 3:972

[81]TAED QP001:10

[82]Paraffin was first identified and named by Karl Reichenbach, the same Reichenbach responsible for the Odic force theory that prompted Edison's own Etheric force theory. W V Farrah, "Reichenbach, Karl (or Carl) Ludwig," in *Dictionary of Scientific Biography*, ed. Charles Coulston Gillispie (New York: Scribner, 1992).

[83]TAEB 3:1099.

[84]TAED QP001:49

Fig. 7.15 Device to scrape wax coated tape to a uniform thickness

Fig. 7.16 Final knife-edge recording point mounted on a brass diaphragm. Wax shavings are visible on the point

on a thin knife edge. This consisted of a 0.4 mm thick steel blade, the point being a 45° V sharpened on two edges, mounted transverse to the tape motion and soldered to the end of the supporting screw (Fig. 7.16).

7.10.6   Finally, a Credible Recording

After many hours of building and experimentation, I achieved something resembling recorded sound. I made an electronic recording of this by holding a microphone over

Fig. 7.17 Grooves cut into the waxed paper tape during recording experiments. Striations on the sides of the groves may represent the recorded sound but may also have been caused by recording point chatter. Offsetting the recording point from the centre of the tape enabled two recordings to be made on each side of the tape; hence, there are two grooves visible

the end of the speaking tube.[85] The recording contains the words "one, two, three, four" and, while the sounds correspond to the pattern to the words spoken, they are not intelligible as words. As such, it fits Edison's comment about one of his early telephone microphones: "if you knew what they were saying it sounded awful like what they were saying".[86]

It had taken me about 70 h to get to this point, even with all of the information available and knowing the outcome. Edison and Batchelor said they achieved recognisable recordings in one night on 18 July 1877, a feat achieved with no precedents and in the midst of work on other projects.

The groove in the waxed tape made by the recording point (Fig. 7.17) resembles the magnified image of a groove in a vinyl record. Despite this similarity, it is possible that the marks visible along the edge of the groove may have been the result of the recording point "chattering" rather than being recorded sound. As the recording point I was using was relatively long (about 20 mm), I speculated that it may have caused chattering so I changed to shorter, stiffer recording points (Fig. 7.16).

7.10.7 The Waxed Tape Phonograph: Mark 2

Having achieved something resembling recorded sound, I decided to build a new device that more closely resembled Edison's and Batchelor's description of their first device (Fig. 7.3), replacing the automatic telegraph style drive with a simple grooved wooden block, mounting the hand Phonograph on a hinged plate over the grove (Fig. 7.18). As this was very close to the instrument in Fig. 7.3, I hoped to get a result

---

[85]Ian Wills. 2011. "Experimental Phonograph Recording, Sample_1." YouTube. https://youtu.be/O4wuFSDngQ8
[86]TAEB 3:767n1 This was on 27 July 1876.

Fig. 7.18 Experimental Phonograph Mark 2

closer to what I understood was Edison's first recording. It produced results no different from the Mark 1 version.

7.10.8 The Waxed Tape Phonograph: Mark 3

Describing their first efforts at recording sound, Batchelor said, "On pulling the paper through a second time, we both of us recognized that we had recorded the speech".[87] Despite many hours of experiment, I could not say that I succeeded in creating a recording recognisable as speech. Listening to the recordings I could tell something was recorded but could, at best, only pick out an occasional word and only then because I knew what I was listening for.

In an effort to achieve recognisable speech, I decided to eliminate a major variable and try something not available to Edison. I replaced the hand Phonograph with a means for electronically generating and reproducing sound. This Mark 3 version still used the recording point and waxed tape from previous experiments but in place of the hand Phonograph and diaphragm, I fitted a small loudspeaker modified to hold the recording point. Figures 7.19 and 7.20 show the Mark 3 device open and closed.

To make a Phonograph recording, I first recorded a few words digitally then played then back through the loudspeaker as I pulled the waxed tape under the recording point. Reproduction used the same process but the vibrations caused by the passing taper were converted into electrical current by the loudspeaker and recorded digitally. This is possible because the type of loudspeaker used generates small electrical currents in response to impressed sound, the same principle as Alexander Graham Bell used in his first Gallows telephone microphone.[88] Bell's

---

[87]TAEB 3:972n4.

[88]Gorman, "Alexander Graham Bell's Path to the Telephone".

Fig. 7.19 The Waxed Tape
Phonograph Mark 3 open

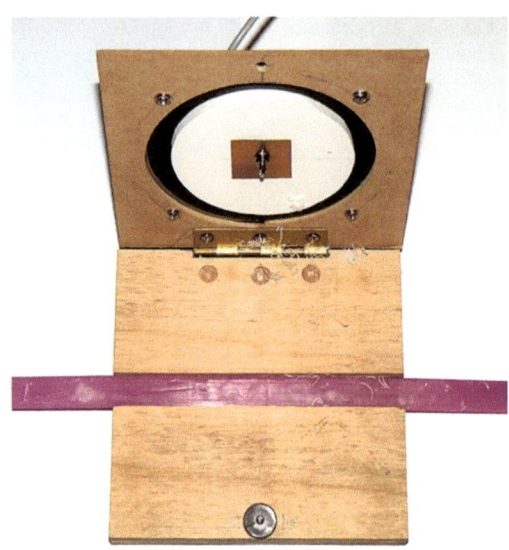

Fig. 7.20 The Waxed Tape
Phonograph Mark 3 closed

device was ineffective for his purpose because of its low power and so could transmit over only small distances. I overcame that limitation with electronic amplification.

The purpose of this radical shift from the Edison-like mechanical device to electronic recording was to remove, as far as possible, variables that might have been adversely affecting recording such as the dexterity required to both pull the tape through and shout into the speaking tube simultaneously. Despite this advance, the result was only marginally better than obtained with my purely mechanical device, just a few scratchy sounds that corresponded to the words spoken. They are still barely intelligible but if the listener knows what was spoken (as Edison and Batchelor would have) it is possible to recognise the words recorded.[89]

It had taken me 115 h of building and experimenting to get to this point whereas Edison and Batchelor conducted their first sound recording experiments over a few hours in one night. Based on my experiments I am convinced they heard no more than I did. In his 18 July notebook entry, Edison wrote, "The new speaking vibrations are indented nicely & there's no doubt that I shall be able to store up & reproduce automatically at any future time the human voice perfectly."[90] From my experiments, it seems that Edison and Batchelor probably heard a pattern of sounds but not intelligible speech and saw a pattern of vibrations recorded in the tape.

7.11 What Edison and Batchelor Heard

In the process of developing the carbon microphone, Edison made numerous laboratory notebook entries describing the kinds of sounds it transmitted and failed to transmit. In contrast, notebook entries relating to the Phonograph before 6 December 1877 contain no such descriptions, just a few vague comments, as when he wrote "OK" next to an experiment on 9 September.[91] The first clear reference to a successful recording is his retrospective account of having recorded "Mary had a little lamb" on 6 December 1877 on the tinfoil Phonograph built by John Kruesi.[92]

This absence of results is significant. By the time he began working on the Phonograph, Edison had many months experience with the carbon microphone and a good understanding of success criteria relating to transmitting the human voice. On 6 July 1877, for example, he wrote that the microphone could get "a good many words Plain such as How do you do" but made no equivalent entries for the Phonograph despite working on both in parallel.[93] It is improbable that he should work on two devices at the same time, both involving articulate sound and test only one with such phrases. The implication of this absence of records of successful recordings before 6 December 1877 is that there were no successful recordings.

We can now connect a number of important pieces of evidence. Firstly, repeating Edison's first experiments indicates that his first recordings were not intelligible although a pattern of sounds corresponding to the words originally spoken would have been audible. Secondly, there is an absence of laboratory notebook entries

---

[89]Ian Wills. 2011. "Experimental Phonograph Recording, Sample_2." YouTube. https://youtu.be/WZ82DUqdCuo; 2011. "Experimental Phonograph Recording, Sample_3." YouTube. https://youtu.be/8eGt8hk8hEw; 2011. "Experimental Phonograph Recording, Sample_4." YouTube. https://youtu.be/e3qYQ0P0kgo

[90]TAEB 3:972

[91]TAED NS77:03.

[92]Quoted in Dyer and Martin, *Edison, His Life and Inventions.* 208.

[93]TAED TI2:34.

describing successful sound recordings and finally, we have Edison's account of his expectations for, and reaction to, Kruesi's first tinfoil Phonograph. His expectation was that he might hear "a word or so" from it, implying that even hearing a few words would have been an improvement on what he had achieved previously. When he heard far more, Edison said he was "never so taken aback in my life".

Taken together, it is reasonable to conclude that that until he heard Kruesi's first tinfoil Phonograph on 6 December 1877, Edison had nothing much better than the first few scratchy sounds he and Batchelor heard on 18 July 1877.[94] Edison's recording of "Mary had a little lamb" on 6 December 1877 is thus history's first credible sound recording, not that of 18 July 1877. Little wonder he rushed to demonstrate it to *Scientific American* the next day.

One thing clear from my attempts to repeat Edison's first experiments is that however little Edison heard on the night of 18 July 1877, the experiment convinced him that he would be able to "to store up and reproduce the human voice perfectly". He came to this conclusion after a few hours work but repeating those first experiments took me over a hundred hours, even with the advantage of knowing how it had been done. Clearly, Edison and Batchelor were either remarkably lucky or they possessed something more. Repeating their experiments suggests that although luck played a part (having a suitable wax at hand for example) Edison and Batchelor were unquestionably very skilful at rapidly creating novel devices. That they could achieve such remarkable results not only on this occasion but also on many others attests to their exceptional ability.

Repeating their crude experiments also shows that even in is this extremely simple device, small variations in detail can mean the difference between getting something and getting nothing. In repeating their experiments, I encountered many problems including finding an effective combination of wax hardness and flexibility, finding and making an effective wax thickness and finding the best recording point shape. People working on novel devices in mature technologies have fewer unknowns because they have developed knowledge that allows better, though not perfect, prediction of what properties materials need and what configuration is likely to work. When working with a novel technology in a completely new field as the Phonograph was for Edison at the end of 1877, that knowledge did not exist, it had to be acquired. Edison's solution, and mine, was to use trial and error to build systems of regularities to predict what was likely to succeed and what was not.

7.12 Edison Pursues Sound Recording and Reproduction

On the night of 18 July 1877, Edison and Batchelor did not hear intelligible recorded speech, despite their subsequent claims. In the light of this we can re-examine a statement made by Batchelor about their first experiment in which he said "On

---

[94]Ian Wills. 2011. "Experimental Phonograph Recording, Sample_1." YouTube. https://youtu.be/O4wuFSDngQ8

pulling the paper through a second time, we both of us recognized that we had recorded the speech".[95] While on first reading it might seem that he meant that they heard the words recorded, Batchelor does not say this explicitly, only that they recognised that they had recorded speech. Being able to hear a pattern of sounds that corresponded to the original words could also fit his description of recognising that they had recorded speech.

If they heard so little in this first experiment, what convinced Edison that he would be able to store up and reproduce sound perfectly? In part, the answer is Edison's enormous optimism and confidence in his own ability. In 1875, on similarly limited evidence, he optimistically declared that he had discovered a new force of nature, Etheric force. This optimism was not limited to his Etheric force theory; it was common for Edison and crucial to his success, and at times failure, as an inventor. A significant proportion of Edison's patent applications were based on the kind of scant evidence that initiated his work on the Phonograph. It may have been scant but it was sufficient to support Edison's belief that these inventions were possible.

At times, this optimism and self-confidence failed Edison. Examination of Edison's 1084 utility patents suggests that many came from misguided beliefs and consequently did not become commercially successful products. Edison's fame as an inventor may rest on his 1084 patents, but his success as an inventor and entrepreneur came from just a few but crucial inventions and patents. He succeeded commercially, in part because he produced so many patents, only a small proportion of which needed to be commercially successful for Edison to be successful overall as an inventor. This is not to suggest that Edison's success came from applying for many patents. Rather, that Edison could recognise a potential invention from very limited evidence and had the skill and resources to develop it.

His optimism meant not only that he pursued meagre evidence to viable inventions; it also meant that he was not discouraged by failures, instead using them to move towards success. Despite this, optimism on its own was not enough to yield worthwhile inventions. It needed some viable basis, no matter how slight that may have been. Without this, optimism would have been no more than wishful thinking. If Edison did not hear intelligible recorded speech on 18 July, what then was that viable basis that led him to believe he would be able to record and reproduce sound?

The evidence of Edison's first notebook entry and recreation of his first experiments points to two likely candidates. The first is the hand Phonograph, which is so simple and direct that even a partially deaf man like Edison could detect vibration caused by sounds. Batchelor then incorporated this device into their first wax tape recording experiments. Examination of the recorded tape showed that the vibrations Edison felt with the hand Phonograph were "indented nicely" in something permanent, marks on the side of the groove cut into the waxed paper tape. That is, Edison saw something like the effect shown in Fig. 7.17.[96] In the light of this, the fact that the first sounds recorded were probably unintelligible was less significant to Edison's

---

[95]TAEB 3:972n4.
[96]TAEB 3:972

optimism than the connection between feeling the vibrations from his voice with the hand Phonograph and seeing them apparently recorded on the tape.

7.13 Why Edison?

One of the most notable features of the first successful Phonographs (Fig. 7.1 and Fig. 4.1) is their simplicity. They consist of a simple assembly of well-known components; the screw, crank, metal foil and inscribing point; components that had been in use for thousands of years. Moreover, unlike the carbon microphone and electric light that depended on nineteenth century electrical technology, these first Phonographs employed none of the electrical wizardry or exotic chemical process that established Edison's reputation as an ingenious inventor. Given their simplicity (far simpler than medieval cathedral clocks, for example), there nothing in the device that would have prevented the Phonograph from having been invented centuries before Edison. Edison's initial motivation for inventing a means for recording sound was an immediate need to record telephone messages but had the Phonograph been invented in the centuries before Edison and the telegraph, there would undoubtedly have been demand for it.

Why then, was it Edison, and not someone centuries before him, who invented the Phonograph? Key to this is that Edison recognised that it was possible to transfer the vibrations felt with the hand Phonograph onto a recording medium. Secondly, he and Batchelor were able to put together their first crude apparatus using components that were available to them: a telephone mouthpiece, some waxed paper tape and a grooved block from the recording telegraph. Third, as repeating their experiments has shown, working together Edison and Batchelor possessed remarkable skill and ingenuity in rapidly creating novel devices. Even before the Phonograph, Joseph Henry had praised Edison as "the most ingenious inventor in this country … or in any other".)[97] The evidence of the Phonograph supports Henry's opinion, but so did Edison's record of inventions. By July 1877, Edison at the age of 30 already held over 100 patents and had a string of important inventions to his name. This ingenuity was supported by the work of two other remarkable men, Charles Batchelor and John Kruesi. In 1879, a newspaper said that Batchelor was "so intimately associated with [Edison] in his work that [Batchelor's] absence from the laboratory is invariably a signal for Mr. Edison to suspend labour".[98] Kruesi did not work beside Edison as Batchelor did but his skill was crucial. He translated Edison's crude Phonograph sketch (Fig. 7.8) into the successful prototype Edison demonstrated to *Scientific American* (Fig. 7.1). "If the devices that emerged [from Kruesi's workshop] didn't work, it was because they were bad ideas, not because they were badly made. And

---

[97]TAEB 3:1117.
[98]New York Herald, December 21, 1879 quoted in Jehl 1937, 393.

when the ideas were good, as in the case of the Phonograph, the product of Kruesi's shop would prove it".[99]

To a degree there was an element of good fortune in Edison's first experiments, most importantly, having waxed paper tape that was neither too brittle, too soft nor the coating too thin to make a recording. In 1904, he told a new employee, "I do not believe in luck at all. And if there is such a thing as luck, then I must be the most unlucky fellow in the world. I've never once made a lucky strike in all my life"[100] Edison may have believed this in 1904, but in 1878 he acknowledged its role, describing his work on electric lighting as requiring "hard work and some good luck".[101] Edison did not just hope for good luck, he helped it, in this instance by having a very well stocked material store to draw from.

Individually, none of these factors on their own would have been decisive. Brought together on that night in July 1877, they yielded success. That they came together in that time and place was no coincidence. Creating of inventions like this was Edison's express objective when he set up his Menlo Park complex, bringing the right people and resources together in an invention-focused environment.

One further element contributed to the Phonograph demonstrated to *Scientific American*; Edison's crucial change in the function of the Phonograph. Initially, he intended it to be a means for recording telephone messages as part of the telegraph system, an important component, but of itself no more significant than many of his other telegraph inventions. Edison dabbled in experiments related to this kind of invention regularly while working on larger projects, as his notebook entries between July and November 1877 indicate. However, on 23 November, he discussed the potential of the Phonograph for home entertainment using mass-produced recordings.[102] Having recognised that it had much greater potential than in the telegraph office, Edison's reason to develop it into a viable invention increased substantially. It was no longer just a useful device for the telegraph office, it was potentially the basis of a new mass entertainment industry, whose products were not just Phonographs but the recordings played on them.

7.14   Some Reflections on Experimental History of Technology

At this point, it is worth reflecting on experimental history of technology more generally. The first and obvious reflection is that I am not Thomas Edison, nor did I have a Charles Batchelor at my side. They were professional inventors, in the business of getting results rapidly. I am not. Having said that, I do not think the

---

[99]Friedel and Israel 1987, 35.

[100]Quoted in Rosanoff, "Edison in His Laboratory," 406.

[101]TAEB 4:1570.

[102]TAEB 3:1119.

time I spent in trying to achieve what they said they had achieved implies a significant lack of ability on my part. Rather, in hindsight it appears that much of my time was spent searching for something that was never going to be achievable. The evidence of my experiments is that paper tape coated with the kind of waxes Edison and Batchelor used in the kind of apparatus they described cannot be is not suitable for making intelligible recordings of speech. Not being Edison and Batchelor, I was unlikely to have achieved their results in a few hours over one night. However, spending more than 100 h without achieving a readily recognisable recording is convincing evidence that it was not achievable with their first materials and equipment.

The second reflection relates to the correspondence between the original experiment and its recreation. I did not attempt to copy Edison's experiment in detail. To do so would have been extremely difficult, if not impossible, given the incomplete records available. Edison used part of an automatic telegraph, which I did not have. However, it was not its automatic telegraph function he made use of, but its ability to move and guide paper tape. That function is achievable without an original automatic telegraph. The aim of experimental history of technology is a functional recreation of the original, not a copy of it, especially when there is not enough historic information to do so confidently.

Related to this is a third reflection: it is possible to apply knowledge developed later to determine what is important and what is not for an accurate functional recreation. Edison's hand Phonograph was adapted from a telephone mouthpiece he was working on at the time. It was probably a different shape from mine and tube made from different material (compare Figs. 7.11 and 7.13). Current theoretical understanding indicates that its shape and material would have had negligible effect on the vibrations felt in the diaphragm so the difference is of no consequence. On the other hand, there is good reason to believe that the nature of the waxed tape and recording point were critical, so I put considerable effort into experimenting with these.

A far more radical shift is embodied the Mark 3 waxed paper Phonograph, which used components and technologies unavailable to Edison. Despite this, the crucial parts, the recording point and waxed paper tape, were as close to Edison's original as can be determined from the available primary sources. This radical change in technology does not negate the conclusions drawn from the experiment because all that has effectively changed is the means for making the recording point vibrate when recording and sensing its vibration in reproduction.

Discussing a similar issue, Sibum, who repeated Joule's mechanical equivalence of heat experiments, made this observation:

> When scientists, historians, sociologists and philosophers aim at exploring the margins of experiment, an important issue is replication, which should not be thought to signify (as the word might suggest) that the experiment concerned must be identical to the original in its construction, and in the materials, instruments and practices employed. By contrast, an experiment differing in one of these parameters is regarded to provide an even more

convincing replication, if it only represents the phenomena credibly or is able to reproduce the measured values exactly.[103]

My fourth reflection relates to success and failure. When I started to recreate Edison's experiments I believed, based on my understanding of statements he and Batchelor made, that they heard a the human voice recorded in their first experiment on 18 July 1877 and in subsequent experiments between then and December 1877. I pursued this objective over many hours but ultimately I failed to achieve an intelligible recording. Despite being a failure in terms of my original objective of making an intelligible recording, it is a success because it shows that Edison and Batchelor exaggerated their claims for their early sound recording experiments. More importantly, it highlights Edison's ability to seize weak phenomena and believe they could be developed into viable inventions. Edison took some weak vibrations felt in the diaphragm of his hand Phonograph, their apparent indentation in waxed tape and the scratchy sound it produced, as evidence that he would be able invent something that could "store up & reproduce automatically at any future time the human voice perfectly". Doing this required enormous self-confidence and in this instance, Edison was justified.

7.15   Innovation

The Phonograph was significant in Edison's life but it is also important in illustrating the earliest stages of innovation. Recognising that something is possible (in this case, the ability to record and reproduce sounds) is not the end of the process, but the beginning. Developing the invention firstly into something that can be patented, then turning it into a commercial product, involves taking the possibility that has been recognised, finding the most effective combination of components (particularly the materials from which they are made), and working then into an effective configuration.

In the case of the Phonograph, Edison first identified a function to be fulfilled, and then found some scraps of evidence that it could be achieved. Based on those scraps of evidence he made the bold declaration that there was "no doubt" he would be able to record and reproduce sounds "perfectly". That weak evidence and strong conviction sustained him through months of what now appear to have been largely failure. As he said himself, it needed only a little inspiration but a lot of perspiration.

Crucial to the process of innovation illustrated by the Phonograph is the difference between the initial recognition that the inventor had a viable means of achieving the desired function and the process of development, first to a patentable form, then to a practical device to be sold to others. Invention is not just that first viable

---

[103]Otto Sibum, "Experimental History of Science," in *Museums of Modern Science: Nobel Symposium 112*, ed. Svante Lindqvist, Marika Hedin, and Ulf Larsson (Canton, MA: Science History Publications/USA, 2000), 77.

experiment. Edison began work on the incandescent lamp in July 1878 but did not apply for the controlling patent until November 1879. Despite this gap of 16 months, the more significant gap is that between Edison's first lamp that reached 500 h of operation in January 1880 and its origins, Humphrey Davey's observation that a wire carrying electrical current glowed incandescent for a few seconds in 1812.[104] Davey may have been the first to observe the phenomenon but 68 years and dozens of inventors separate the two events. As the Phonograph and incandescent lamp demonstrate, there is far more to inventing than recognising a phenomenon that is later developed into an invention. It is mistaken to equate discovery with invention.

---

[104]Davy, *Elements of Chemical Philosophy Part I Vol I*, 85.

Chapter 8
Scientific Failure: Etheric Force

8.1 Edison Patents a Wireless Communication System

In 1885, Thomas Edison added to his already substantial tally of 141 patents in telegraphy, a patent for sending telegraph messages without wires. This invention, he claimed, would enable communication "between distant points ... by induction without the use of wires connecting such distant points" (Fig. 8.1). Despite Edison's patent predating Heinrich Hertz's electromagnetic radiation experiments by 3 years and Guglielmo Marconi's wireless telegraphy by ten, Edison did nothing with the invention.[1] Given Edison's ability to make much of very little, as he did with the Phonograph, it was uncharacteristic that he did not exploit this revolutionary invention and become known as the inventor of wireless telegraphy. Instead of developing it, Edison sold the patent rights to Marconi. The reason for Edison's ambivalence is to be found in a controversial incident 10 years earlier.

In November 1875, Edison announced to the press that he had discovered an "entirely unknown force [of nature], subject to laws different from those of heat, light, electricity or magnetism".[2] His discovery, which he named Etheric force, sparked public interest but also prompted doubters in the scientific community. Edison experimented with the phenomenon for about five weeks, his venture into public scientific theory-making ending the following July when he privately abandoned his Etheric force theory and accepted his opponents' explanation that the phenomenon was no more than a form of electrical induction. Not only did Edison

---

Some of this chapter appeared previously in Ian Wills, "Edison and Science: A Curious Result," *Studies in History and Philosophy of Science Part A* 40, no. June (2009).

---

[1]Heinrich Hertz (1857–1894). Hertz studied engineering at the Dresden Polytechnic and later physics under Helmholtz and Kirchoff. He was awarded a PhD in 1880 and in 1885, he was appointed to the University of Karlsruhe where he worked with electromagnetic radiation, providing experimental verification of Maxwell's theories. Hertz died of bone malignancy aged 36.
[2]TAEB 2:678.

© Springer Nature Switzerland AG 2019
I. Wills, *Thomas Edison: Success and Innovation through Failure*, Studies in History and Philosophy of Science 52, https://doi.org/10.1007/978-3-030-29940-8_8

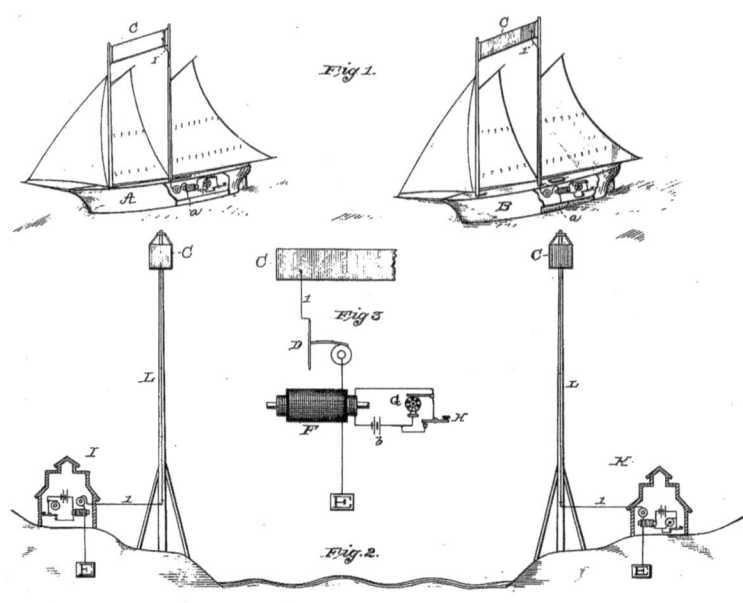

Fig. 8.1 Edison's patent for wireless telegraphy (Thomas A Edison. Means for Transmitting Signals Electrically. US Patent 465,971, filed 23 May, 1885, and issued 29 December, 1891)

fail to have his Etheric force theory accepted but in accepting his opponents' explanation he failed to exploit a phenomenon that fell within an area in which he was an acknowledged expert, electrical technologies. Edison's Etheric force was neither a new force of nature nor an induction effect as his opponents asserted but caused by electromagnetic radiation, the basis of wireless communication.

The course of Edison's Etheric research and the associated controversy illustrates aspects of the relationship between the inventors' methods and those of scientists. Looking at Edison's and his opponents' approaches to Etheric force shows that the inventor's approach to creating a successful invention is identical in key areas to that used by scientists to develop successful theories. Edison failed to have his Etheric force theory accepted because of his beliefs about scientific research and the development of theories. He failed not because he was an inventor who did science badly, but because he was an inventor who, when he turned to developing a scientific theory, abandoned techniques and perspectives he used so successfully in producing inventions.

As an inventor, Edison was primarily in the business of creating artefacts, things made to serve a purpose or function. Many of the artefacts Edison created were physical inventions, like the carbon microphone, incandescent lamp and the Phonograph but his success as an inventor also required him to create non-physical artefacts. These non-physical artefacts included patent applications and companies to exploit his inventions. Both physical and non-physical artefacts are the product of art in the older sense of the word, in which art means skill in doing something, skill that comes from acquired knowledge. Art in this sense is recognised in patent laws which require that a patent not be the product of "prior art", that is existing public knowledge.[3]

While the Etheric force controversy may have been a contest of scientific theories, it was also a contest of artefacts in which Edison, the iconic inventor, the creator of ingenious artefacts, was beaten by other inventor-scientists who created more successful artefacts. It is irrelevant years later both Edison's Etheric force theory and his opponents' induction theory were judged erroneous. In 1875–76, the induction theory was more successful because it was more convincing. While we do not normally think of science and the work of scientists as directed towards making artefacts, the Etheric force controversy demonstrates the key role artefacts play in the acceptance of theories.

8.2   The Etheric Force Debate

Edison's search for a new force of nature began within days of meeting Dr. George Miller Beard, a New York physician.[4] Edison and Beard met in connection with one of Edison's inventions, the Inductorium which used induction coils to deliver electric shocks, Edison's advertisements claiming, "This instrument should be in every family as a specific cure for rheumatism and as an inexhaustible fount of amusement".[5] While electric shocks were an established therapy in the nineteenth century, the "inexhaustible fount of amusement" came from delivering shocks to unsuspecting people for the amusement of initiated onlookers, Edison being fond of practical jokes involving electric shocks.[6]

It was the therapeutic properties of the Inductorium that interested Beard who an influential figure in late nineteenth century psychiatry and had just published the second edition of his book on the medical and surgical uses of electricity.[7] As a pioneer of electrotherapies, Beard hoped to use Edison's Inductorium to treat Neurasthenia and other nervous disorders.

---

[3]US Department of Commerce, *Manual of Patent Examining Procedure*.

[4]George Miller Beard (1839–1883) Beard was an influential figure in late nineteenth century psychiatry and pioneer of electrotherapies. He published several editions of a textbook on the medical and surgical uses of electricity and edited the journal, *Archives of Electrology & Neurology*. Beard introduced the term Neurasthenia and later published *American Nervousness, Its Causes and Consequences* (George Miller Beard, *American Nervousness Its Causes and Consequences* (New York, NY: G. P. Putnam's Sons, 1881). Beard's book was read by, and influenced, Freud. (Philip P Weiner, "G M Beard and Freud on 'American Nervousness'," *Journal of the History of Ideas* 17, no. 2 (1956).)

[5]TAEB 2:434, 2:435.

[6]Conot, *A Streak of Luck*, 25–26.

[7]George Miller Beard and Alphonso David Rockwell, *A Practical Treatise on the Medical & Surgical Uses of Electricity: Including Localized and General Faradization; Localized and Central Galvanization; Electrolysis and Galvano-Cautery* (New York: William Wood & Co, 1875).

Apart from his professional interests, Beard was also a crusading proponent of science and opponent of supernatural explanations of phenomena. Around the time of their meeting, Beard was engaged in a bitter public debate with Helena Blavatsky, spiritualist and founder of the Theosophical Society.[8,9] Beard's skirmish with Blavatsky followed his public claim that two of her favoured psychic mediums were frauds. Edison later had considerable contact with Blavatsky and her deputy, Henry Olcott, joining the Theosophical Society in 1878 and giving Blavatsky one of his earliest Phonographs.[10,11]

A few days after his introduction to Beard, Edison made the first notebook entry describing a series of experiments on "Odic magnetism".[12] Given the publicity about Beard's opposition to spiritualism, it is likely that Edison and Beard discussed the Odic force theories of the German chemist and mineralogist, Karl von Reichenbach.[13] A resurgence of interest in mesmerism and animal magnetism in the 1840s led Reichenbach to investigate the effect of magnets and crystals on humans. Based on these experiments, Reichenbach claimed to have discovered Odic force (or Od), a mysterious natural force he believed pervaded all things and could explain phenomena as diverse as the Aurora Borealis, clairvoyance and ghosts.

Reichenbach's works were translated into English and became popular in the United States after publication of an American edition of his Odic force book and an

---

[8]Marion Meade, *Madame Blavatsky: The Woman Behind the Myth* (New York: G P Putnam's Sons, 1980), 127.

[9]Helena Blavatsky (1831–1891). Russian born spiritualist and co-founder, with Henry Olcott, of the Theosophical Society in American. Blavatsky translated an article on Edison's Phonograph into Russian for the Russian newspaper, *Pravda*.

[10]Henry Steel Olcott (1832–1907) Olcott was Civil War commander and lawyer who became a follower of the spiritualist, Helena Blavatsky. With Blavatsky, Olcott founded the Theosophical Society in America and, on Blavatsky's death in 1891, succeeded her as head of the Society.

[11]TAED D7802:1, TAED D8912:1–3.

[12]TAED NS7401:43–46.

[13]Karl Reichenbach (1788–1869) Karl Ludwig Freiherr von Reichenbach was a German scientist, inventor and industrialist. He received a doctor of philosophy from the University of Tübingen and became involved in iron and steel production, amassing a significant fortune and the title of baron (Freiherr). Reichenbach was an early researcher in organic chemistry, discovering creosote, waxy paraffin, phenol (carbolic acid) and pittacal, the first synthetic dye. In geology, Reichenbach produced the first geological survey of Moravia and the region around Prague and published papers on the mineralogy of meteorites. His interest in meteorites led him to assemble a large private collection of specimens that he bequeathed to the University of Tübingen. In the 1840s, Reichenbach turned to investigating the effect of magnets and crystals on humans leading him to claim to have discovered Odic force (or Od). His interest in chemistry also led him to experiment with photography, a revolutionary technique in this period, and to employ it in some Odic force experiments.

article on it in the *American Whig Review*.[14,15] The influence of his ideas spread widely, with many literary references to Odic force and mesmerism, including some in the works of Edgar Alan Poe.[16] Given the publicity his ideas received, it is likely that Edison knew something of Reichenbach and his Odic force before he met Beard. Although Reichenbach's theories received some scientific support initially, this support effectively ended in 1862 when seven Berlin professors, including the physicist Heinrich Magnus, published a letter repudiating his Odic theories.[17]

Like Beard, Reichenbach expressed a strong belief in the superiority of science over superstition and saw his work on Odic force as dispelling superstitious beliefs by replacing them with rational scientific explanations. It is somewhat paradoxical then, that his work was dismissed by scientists but embraced by spiritualists like Blavatsky.[18] Given Beard's opposition to spiritualism and his own questionable grasp of scientific method, it is not surprising that he should be interested in Reichenbach's theories.

Scepticism of Reichenbach's theories among professional scientists (Beard was an exception) did not deter Edison, who conducted more experiments in search of "a new force" in May and June 1875. On 31 May 1875, Edison drew up a list of potential research topics including "A New force for Telegraphic communication".[19] The search for a new force was clearly on Edison's agenda.

During the night of 22 November 1875, while experimenting with a new signal generating vibrator for the acoustic telegraph (Fig. 8.2), Edison and his associate, Charles Batchelor, noticed sparks at S, a point at which no current should have been flowing.[20] On investigating further, they found that they could also draw sparks from other parts of the vibrating bar R and from the end of a wire connected to X. When they connected the wire to a metal gas pipe, they found they could draw sparks throughout the room by touching metal objects to other parts of the gas pipe. To their amazement, they found they could even get sparks by bending the wire into a loop and touching it back onto itself, something regarded as impossible within the prevailing direct current and static electric theories of the period.

Although they had seen similar sparks often before and had previously attributed them to electrical induction, Batchelor commented that these sparks "seemed so

---

[14]Baron Karl von Reichenbach, *Physico-Physiological Researches in the Dynamic of Magnetism Electricity Heat Light, Crystallization, and Chemism in Their Relations to the Vital Force*, trans. Leslie O Korth, 2nd American ed. (New York: Partridge Britton, 1853).

[15]The American Whig Review, "Researches of Baron Reichenbach on the "Mesmeric," Now Called the Odic Force," *The American Whig review* 15, no. 90 (1852).

[16]Martin Willis and Catherine Wynne, eds., *Victorian Literary Mesmerism*, Costerus, New Series., V. 160. (Amsterdam: Rodopi, 2006).

[17]Farrah, "Reichenbach, Karl (or Carl) Ludwig."

[18]See, for example, H P Blavatsky, *Isis Unveiled: A Master-Key to the Mysteries of Ancient and Modern Science and Theology*, vol. 1 (New York: J W Bouton, 1877), 125. http://www.theosociety. org/pasadena/isis/iu-hp.htm

[19]TAEB 2:570, 2:579, 2:581.

[20]TAEB 2:665.

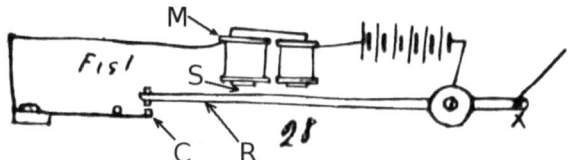

Fig. 8.2 The experimental vibrator (notation added) on which Edison first noticed Etheric force sparks (TAED NE1691:15). When the contact C closes, it completes the circuit, causing the electromagnet M to pull the iron rod, R towards S. This breaks the circuit by opening the contact C and so the rod to returns to its original position. The process is cyclic causing the rod to vibrate. In operating principle, it is identical to an electric bell, in which this repeated motion causes a hammer attached to R to repeatedly strike a gong

strong that it struck us forcibly there might be something more than induction".[21] Edison went further and declared from this startling but limited evidence that "This is simply wonderful & a good proof that the cause of the spark is a true unknown force".[22] This might seem to be an extraordinarily grand claim to make on such limited evidence but it was not spontaneous. As Israel observes, the reason why Edison thought he had found a new force was simply that he had been searching for one for the past year.[23]

Encouraged by his apparent success, Edison and Batchelor experimented during the following nights. Less than a week later, Edison announced his discovery to the press with newspaper reports beginning to appear on 29 November 1875. Most reports were positive, describing Edison's announcement as a "Wonderful Invention" and a "Startling Discovery", and proclaiming that it would lead to a new era in communication.[24] The *New York Herald* carried a lengthy article on Edison's discovery that included his prediction that it would put an end to "The cumbersome appliances of transmitting ordinary electricity, such as telegraph poles, insulating knobs, cable sheathings", resulting in "a great saving of time and labour". The article also summarised Edison's Etheric force theory. Noting that it was already common knowledge that heat, electricity and magnetism could be converted into each other, Edison continued:

> It follows that if electric energy under certain conditions is transformed into that of magnetism under other conditions it might be transformed into an entirely unknown force, subject to laws different from those of heat, light, electricity or magnetism. There is every reason to suppose that Etheric energy is this new form. The only manifestation of its presence previously recorded with scientific accuracy is that of the German chemist Ruchenbach [sic] ... This phenomenon, inexplicable to Ruchenbach, is easily to be accounted for on the Etheric theory.[25]

---

[21]TAED MBN002:4.

[22]TAEB 2:665.

[23]Israel, *Edison: A Life of Invention*, 111.

[24]TAEB 2:678n3, n5.

[25]TAEB 2:678.

Edison's desire to discover new a new force of nature and to show the interrelatedness of natural forces reflected his admiration for the work of Michael Faraday but his reference to Ruchenbach (Reichenbach) was to prove unwise. In one of the few negative reports on Edison's claims, the *New York Times* seized on Edison's mention of Ruchenbach, described him as the "maligned and discredited Reichenbach", emphasising the connection between Odic force and "supernatural wonders" such as clairvoyance. It went on to lampoon Edison's gas pipe demonstrations, concluding with the ironic observation that Edison was wasting his time with gas pipes and should instead "begin the manufacture of ghosts and establish direct communication with the other world".[26] (The *New York Times* was unintentionally prophetic in this. Four decades later, Edison was reported as working on a machine to do exactly that: communicate with the dead.[27]).

Between 22 November and the middle of December, Edison attacked Etheric force with his characteristic enthusiasm. Once a problem had his attention, it tended to consume him to the exclusion of all else, sleep, home life and personal comfort included. He and his associates worked on Etheric force night and day, conducting an extensive array of experiments. One strand of his experimental program was directed towards excluding electricity as possible sources of the sparks, Edison concluding that "these sparks or force ... do not follow the laws of either voltaic or Static electricity".[28] (The reason why Edison did not detect electricity was that he was dealing with high frequency alternating current and his tests detected only to direct current and static electricity.)

Another set of tests sought to eliminate electrical induction as the cause since, as Batchelor noted on 22 November, in the past they had attributed similar sparks to induction. One of these tests involved removing the iron core from the electromagnet (M in Fig. 8.2), Edison noting that he still got "sparks just the same".[29] (Although Edison believed he had eliminated the magnetic fields by removing the iron cores, the wire spools were also inductances.) Other sets of experiments involved altering the circuit and adding components to it while observing the effect on the Etheric force sparks.[30]

The largest group of tests involved applying Etheric force sparks to various metals, liquid solutions and powders to determine their effect.[31] Edison had patented a number of inventions that exploited electrically initiated chemical reactions including the recording telegraph he took to England in 1873 and later an electrochemical telephone receiver, the basis for Edison's "musical" telephone and the

---

[26]New York Times. "Etheric Force." *New York Times*, 3 December 1875, 4.

[27]B C Forbes, "Edison Working on How to Communicate with the Next World," *American Magazine* XC, no. 10 (1920).

[28]TAED NE1691:15.

[29]TAED NE1691:18.

[30]TAEB 2:666–2:669.

[31]TAEB 2:666, 2:669, 2:673, 2:680.

"loud-speaking" telephone receiver.[32,33] These telephone receiver exploited his 1874 discovery of a phenomenon that he referred to as the electromotograph principle. Edison seems to have hoped that this series of experiments would identify new phenomena to exploit in other electrochemical inventions.

A major interest of Edison was the potential application of Etheric force to communications. To this end, on 24 November, he connected the Etheric force apparatus to a telegraph line running from his Newark, New Jersey, laboratory to New York, and back. When he found he could draw sparks from the return end of this wire Edison concluded that "This force can be transmitted over long telegraph wires [and] may be transmitted over uninsulated iron wires buried in the earth for instance the sheathing of the Atlantic Cable".[34] This entry, like many other records of the Etheric force experiments, is in Batchelor's handwriting. In recording it, Batchelor appears to have been acting as Edison's secretary for he also kept his own notes that, in this instance, contradicted the official laboratory notebook entry. Privately, Batchelor wrote, "it might be that the force travels across the table instead of going out on the line".[35] It was a crucial speculation. If it was correct and the signal crossed the table without a conducting medium, it was evidence of wireless transmission.

The result of the Newark to New York telegraph experiments was just one of many remarkable characteristics of the new force. On 30 November, Edison found that by holding the gas pipe in one hand he could draw sparks from metal objects using a metal rod held in the other "showing that the force passed through his body".[36] Batchelor's private notebook includes a further experiment in which they achieved the same result with three people holding hands in a chain.[37] (As with the Newark to New York telegraph experiment, this was due to wireless transmission, not conduction through humans.)

During this period of intense experimentation, Edison undertook some development of his Etheric force theory beyond that announced to the press at the end of November. Even so, the theory remained limited. Edison's fascination with particular aspects of Etheric force phenomena, especially the appearance of conduction, points to critical limitations of his theory. The first is that his experiments concentrated primarily on the narrow objective of building systems of regularities (repeatable patterns). The second limitation was more fundamental because it lay in Edison's assumptions about the character of Etheric force. Up to this point in his career, Edison had primarily worked in the electrical field and conceived Etheric force in terms with which he was familiar: direct current, static electricity and, most

---

[32]Thomas A Edison. Paper for Chemical Telegraphs Etc. US Patent 132,455, filed 16 April, 1872, and issued 22 October, 1872.

[33]Electro-Chemical Receiving Telephone.

[34]TAED NE1691:17.

[35]TAED MBN002:4.

[36]TAEB 2:673.

[37]TAED MBN002:6.

significantly, electrical conduction. This latter assumption is apparent in the conclusion he drew from the Newark to New York telegraph line experiment, and in his belief that Etheric force passed through the human body.

Further, since he mentioned Reichenbach in his announcement to the press, Reichenbach's Odic force appears also to have influenced his Etheric force theory and some of his experiments. Like Edison, Reichenbach also conceived Odic force in terms of conduction, claiming that Odic force "is conductible through all other bodies; it is capable of being either directly accumulated on, or transferred by distribution to other bodies".[38]

On 3 December George Beard, the man who seems to have stirred Edison's interest in Reichenbach's Odic force, visited Edison's laboratory with another professional inventor, John E Smith. Smith expressed the opinion, shared by many others, that Etheric force was merely a consequence of electrical induction. Beard brought live frogs, which they killed, testing the effect of Etheric force sparks on the frogs' legs, as Luigi Galvani had done. The frogs' legs were seen to twitch in response to static electric impulses but not when Etheric force was applied, further confirming to Edison that Etheric force was not a form of static electricity.

When they left, Smith and Beard took with them a diagram of Edison's apparatus and used it to experiment independently on Etheric force.[39] Beard later published a favourable account of his own experiments and those witnessed at Edison's laboratory.[40] He also energetically supported Edison's Etheric force theory through his own journal, *Archives of Electrology & Neurology* and wrote to others countering its critics.[41]

Beard's was a minority view so he sought support from other scientists. On 11 December, Beard sent an account of his experiments on Etheric force to a mutual acquaintance, George Barker, professor of physics at the University of Pennsylvania and at one time president of the American Association for the Advancement of Science.[42,43] Barker showed little interest at the time and did not visit Edison until later in the month when he also expressed the opinion that Etheric force was merely an induction effect. Barker wrote, "It seems to me clear that the 'force' is only an inductio[n] current. I hope to settle the question this week".[44] While Barker became a strong supporter of Edison in later years, at this point he did not support his Etheric

[38]Reichenbach, *Physico-Physiological Researches*, 116.

[39]TAEB 2:679n5.

[40]George M Beard, "The Newly-Discovered Force," *The Quarterly Journal of Science, and Annals of Mining, Metallurgy, Engineering, Industrial Arts, Manufactures and Technology*, no. April (1876).

[41]Baldwin, *Edison: Inventing the Century*, 65.

[42]TAED X107:1.

[43]George Frederick Barker (1835–1910) Barker was professor of physics at University of Pennsylvania and at one time president of the American Association for the Advancement of Science. In 1879, he invited Edison to join him on trip to Wyoming to observe solar eclipse. During the trip, barker and Edison discussed the possibility of Edison inventing electric lighting. Barker was a key supporter of Edison in the scientific community.

[44]TAED X120B:1.

force theory, leaving Beard as Edison's main scientific ally. Even Beard's support was of questionable value as some of his peers suspected him of being a scientific humbug. A review of Beard book in *The American Journal of Medical Sciences* questioned the safety and efficacy of his treatments, cast doubts on his grasp of scientific practice and described his conclusions as "fallacious".[45] The scientific credibility of Edison's only scientific ally was itself in question.

There is no record of Edison's reaction to public criticisms such as those in the *New York Times* but Etheric force soon drew attention from a quarter that he could not ignore. On 10 December Edison's agent, Norman Miller, wrote inviting him to a meeting with William Orton, president of Western Union. Miller's letter concluded:

> I think that you had better bring in a Statement of expenditures and such vouchers as you have ready, also drawings, etc., and anything that shows work done and progress made. The papers are so full of "new force" that I want you to show that it has not taken up too much of your time.[46]

Orton had engaged Edison in mid-1875 to develop alternatives to Elisha Gray's acoustic telegraph system, which was a threat to Western Union's near monopoly of the telegraph. In return for the rights to Edison's inventions, Western Union would provide financial support for his plans for a purpose-built laboratory at Menlo Park. The implication of Miller's letter was that while Western Union might finance Edison to produce inventions that benefited Western Union, it would not finance Edison's pursuit of a controversial new force. Edison must have allayed these concerns, because four days later, on 14 December, he and Western Union signed the agreement.[47]

Edison acted in the spirit of his agreement and, except for one more experiment in 1875, stopped research on Etheric force. On 26 December, Edison's laboratory notebook noted, "an experiment tried tonight gives a curious result". Figure 8.3 is the sketch that accompanied the entry.

The left hand side of the sketch shows the apparatus Edison used on 22 November (Fig. 8.2). B, C, D and E are sheets of tinfoil hung on insulating supports and the object in the lower right is an instrument Edison invented (the Etheriscope) to observe Etheric force sparks. The Etheriscope consisted of a darkened box containing two carbon points separated by a small gap. In the sketch, one of the carbon points is earthed to a gas pipe; the other connected to the tinfoil sheet E. B and E were 100 inches (2.5 m) apart with no wire or other conducting medium between them. Despite this separation, the accompanying entry notes that Edison and Batchelor "received sparks at intervals although insulated by such space".[48]

---

[45] WSWR, "Art. Xxxiv- [Review of] the Medical Use of Electricity, with Special Reference to General Electrization as a Tonic in Neuralgia, Rheumatism, Dyspepsia, Chorea, Paralysis, and Other Affections Associated with General Debility. With Illustrative Cases. By Geo M Beard, M D and a D Rockwell M D. 12 Mo Pp 65. New York William Wood & Co 1867," *The American Journal of Medical Sciences* 55, no. January (1868).

[46] TAEB 2:687.

[47] TAEB 3:891.

[48] TAED NE1691:29.

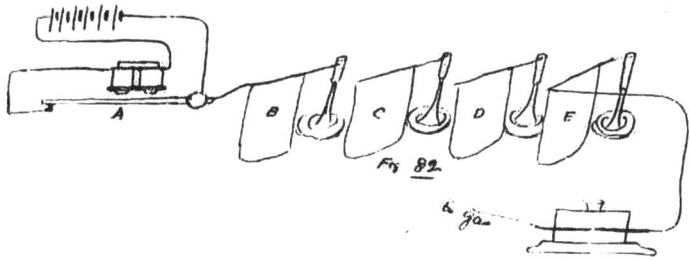

Fig. 8.3 Sketch from Edison's laboratory notebook of the 26 December 1875 showing the wireless transmission experiment (TAED NE1691:29)

This was, indeed, a curious result. What Edison and Batchelor observed with the apparatus was wireless transmission between B and E, confirming Batchelor's speculation of 24 November that Etheric force travelled through space without a conductor. Only the word "curious", hints at the exceptional nature of what they witnessed.

Edison may have stopped his Etheric force experiments in December but the criticism of his theory continued. On 8 January 1876, *Scientific American* published a generally favourable account of the demonstrations by Edison and Beard to the Polytechnic Club of America but in the same and later issues, it also printed letters disputing Edison's theory. On 5 February 1876, *Scientific American* reprinted an article from the *Journal of the Franklin Institute* opposing Edison's Etheric force theory and proposing, as others had done, that the phenomenon could be explained by induction.[49] Its author, Philadelphia Central High School teacher Edwin Houston, disputed Edison's theory and claimed to have reported a similar phenomenon previously. Houston went on to imply that Edison was ignorant of current electrical science.[50,51] A biographer of Houston's partner and fellow Philadelphia Central High School teacher, Elihu Thomson, concludes that Thomson and Houston chose to build their scientific reputations, in part, by attacking the credibility and reputations of others.[52,53] The popular and populist Edison and his Etheric force, linked as

[49]Edwin J Houston, "Phenomena of Induction," *Journal of the Franklin Institute* 101, no. January (1876).

[50]"On a New Connection for the Induction Coil," *Journal of the Franklin Institute* 61, no. July (1871).

[51]Edwin J Houston (1847–1914) American inventor entrepreneur. Initially a Philadelphia Central High School teacher with Elihu Thomson, Houston formed the Thomson Houston Electric Company in 1879. This later merged with the Edison General Electric Company to become General Electric.

[52]W Bernard Carlson, *Innovation as a Social Process: Elihu Thomson and the Rise of General Electric, 1870–1900*, Studies in Economic History and Policy (Cambridge, Cambridgeshire; New York: Cambridge University Press, 1991), 64.

[53]Elihu Thomson (1853–1937) Thomson was a British-born American inventor-entrepreneur with a substantial patent output (696 patents). He was born in Manchester and migrated to Philadelphia in

it was to Reichenbach's marginalised Odic force, presented an opportunity to employ this strategy.

Houston's paper seems to have stung Edison because he responded with a letter to *Scientific American* demanding that his critics "back up their assertions by experiment, and give me an equal chance as a critic".[54] It did not silence Houston, who was capable of more than assertions. Both he and his associate, Elihu Thomson, were competent experimenters and inventors in their own right. In 1879 they established the American Electric Company which later merged with the Edison General Electric Company to become the General Electric Company.

Houston accepted Edison's challenge and, with Thomson, published a more detailed paper in the April *Journal of the Franklin Institute*.[55] As it had done with Houston's first paper, *Scientific American* reprinted this paper in its 20 May issue.

The second paper continued the derisive tone of the first, claiming that Houston and Thomson's explanation, unlike Edison's, was "in accordance with the known laws of electricity", implying that Edison was ignorant of these. The second paper repeated the claim of the first that "all the manifestations classed as 'Etheric' are due solely to inverse currents of induced electricity" but now provided an ingenious demonstration in support. In this experiment, Houston and Thomson split the electromagnet (M in Edison's apparatus in Fig. 8.2) in two, with the wire in the cores wound in opposite directions. They claimed this produced two "charges" of opposite polarity that cancelled, preventing the spark from appearing. Further, their circuit could be adjusted to make the sparks appear and disappear at will. There was no need, Houston and Thomson argued, to resort to Etheric force as an explanation, it was simply an induction effect. (Unknown to Houston and Thomson, the device was a crude tuned radio circuit. The sparks appeared and disappeared as the tuning of the circuit changed.) In describing precautions required for the experiment, Houston and Thomson emphasised the need for symmetry, even of the human operator, in the arrangement of the experimental apparatus. Despite this injunction, they offered no explanation as to why non-conducting, non-magnetic humans could influence electrical induction and reverse currents. (Being a tuned radio circuit, the position of the

---

1858 as a child. On completing schooling at the Philadelphia Central High School, Thomson was appointed a science teacher at the school under his former teacher, Edwin Houston. Despite being a prolific inventor, Thomson preferred to be described as a scientist and throughout his life used the title of professor acquired as a high school teacher. With Houston, he formed the Thomson Houston Electric Company in 1883, which was merged with the Edison General Electric Company in 1892 to form the General Electric Company. Thomson became head of General Electric research after the merger. Houston and Thomson also formed the British Thomson Houston Electric Company and French Thomson Houston Electric Company. The successor to the latter is currently known as Alstom (formerly GEC Alstom), Alstom being a combination of Alsace and Thomson. Given his public clashes with Edison, it is somewhat ironic that Thomson was the first recipient of the IEEE Edison Medal.

[54]TAEB 2:726.

[55]Edwin J Houston and Elihu Thomson, "Electrical Phenomena. The Alleged Etheric Force. Test Experiments as to Its Identity with Induced Electricity," *Journal of the Franklin Institute* 101, no. April (1876).

Fig. 8.4 Edison's
laboratory notebook sketch
of the apparatus he used to
replicate Houston and
Thomson's demonstration
(TAED NS7601:2)

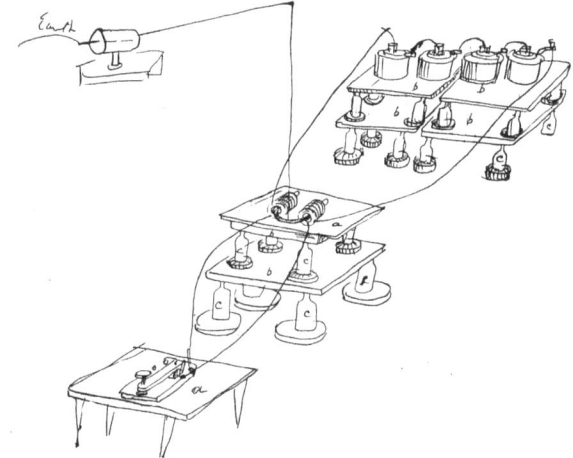

operator altered the tuning in a similar way that a person near the aerial of a radio
may alter its tuning.)

By mid-1876, Edison's enthusiasm for Etheric force had waned under the
combined pressure of opposition from his financial backers, his move to Menlo
Park and his work on other projects, notably the telephone and acoustic telegraphy.
He had not, however, lost all interest in his Etheric force theory and in July 1876 did
as he had threatened, and attempted to take his turn as critic. On 24 July, Edison
replicated Houston and Thomson's demonstration using the apparatus in Fig. 8.4.
Edison went to considerable effort to repeat Houston and Thomson's experiment,
being careful that all parts were well insulated and arranged symmetrically to the
extent that not only the batteries, coils and other components were symmetrical, but
himself. "[I] had my body divided so that equal portions should be on a side.
[I] removed every object from vicinity of magnet that would give a greater amount
of surface or metal on one side or the other = then I closed [the telegraph] key by a
glass rod 3 feet long". He noted that even with all these precautions he still got a
brilliant spark, and initially concluded, "the so called polarity experiments of
Houston & Thompson were incorrectly made".[56]

Despite this initial view, Edison persisted and eventually found that he could,
indeed, make the Etheric force sparks disappear at will, while what he believed to be
their cause, the opening and closing of the circuit, continued. Eventually he con-
ceded, "I think that H & T are confirmed".[57]With this, Edison's excursion into public
scientific theory-making effectively ended.

While his public interest may have ended, privately it lingered for some years and
he experimented with Etheric force on several occasions. In January 1877, Edison

---

[56]TAEB 3:764.
[57]TAED NS7601:14.

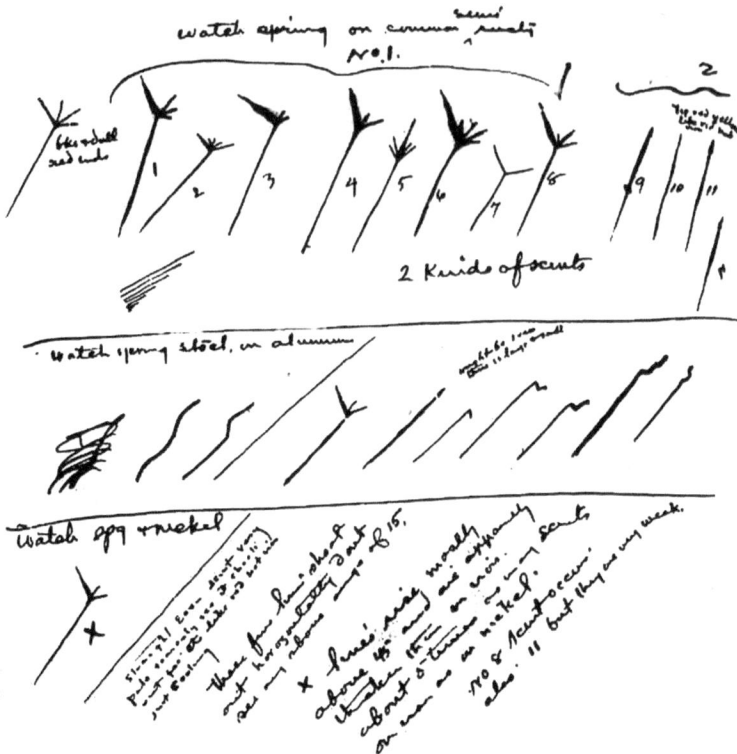

Fig. 8.5 Edison's sketches of Etheric force sparks (TAED NSUN01:2)

and his associates filled more than 100 notebook pages with more experiments testing Etheric force on water solutions, alcohol solutions and acids.[58] Further experiments followed a month later and on 2 August 1877, Edison described and sketched the shape of sparks associated with Etheric force (Fig. 8.5).[59,60]

Even 20 months after his first observation of Etheric force sparks, Edison's experiments remained exploratory. Edison had abandoned his attempt to create theory and instead was seeking phenomena to exploit with an emphasis on chemical effects. Although Edison the scientist seems to have learned that his error was to have made his theoretical claim public, Edison the inventor clung to the notion of Etheric force. On 8 February 1877, he drafted a patent application that began:

---

[58]TAED NV08:1:107.
[59]TAEB 3:864, TAED NV08:177.
[60]TAED NSUN01:2–4.

The object of this invention is to transmit intelligence over or through metal of indefinite length.

The invention consists first in the discovery in physics of a new force or mode of motion, capable of performing work which is neither heat light electricity or magnetism and the application of such discovered force to an instrument capable of indicating its presence to the senses.[61]

This patent application embodies the conclusion Edison came to after the apparent transmission of Etheric force signals between Newark and New York. That is, that Etheric force would enable transmission of signals over long, uninsulated wires.[62] Edison did not complete the draft or apply for a patent. The accompanying sketches show variations on his original Etheric force device (Fig. 8.2) and the Etheriscope.[63] The wording of the patent application indicates that he still thought of Etheric force in terms of conduction over metal wires and had yet not realised that he was working with wireless transmission.

In April 1878, Edison wrote to Beard "Come out and seem me, 'Etheric force' is just as much an unknown mode of motion as it ever was. I am going into it again."[64] Despite this positive tone and a few exploratory experiments, Edison never attacked Etheric force with the enthusiasm he had before his replication of Houston and Thomson's demonstration. Edison had accepted Houston and Thomson's explanation of the phenomenon and abandoned his own Etheric force theory so that, when he turned to work on wireless telegraphy in 1885, it was to exploit induction effects and not Etheric force. In addition to the 1885 wireless transmission patent Edison sold to Marconi (Fig. 8.1) which used induction, Edison took out two other patents for wireless signalling between moving trains and stationary wires (ineffective as it turned out) also described in terms of electrical induction.[65] Houston and Thomson's induction theory had won Edison over.

Edison later regretted his acceptance of Houston and Thomson's explanation when became apparent that Etheric force was a form of electromagnetic radiation and that his 26 December 1875 experiment demonstrated wireless transmission. The loop of wire in Edison's 22 November experiments was effectively a dipole antenna of the kind later used by Hertz to validate Maxwell's theories.[66] After Edison's

---

[61]TAED NSUN01:2.

[62]TAED NE1691:17.

[63]TAED NSUN01:2–4.

[64]TAED X120B:1.

[65]Thomas A Edison. System of Railway Signaling. US Patent 350,234, filed 7 April, 1885, and issued 5 October, 1886; System of Railway Signaling. US Patent 486,634, filed 7 April, 1885, and issued 22 November, 1892.

[66]Heinrich Hertz, *Electric Waves: Being Researches on the Propagation of Electric Action with Finite Velocity through Space* (Reprint, New York: Dover (1962), 1893), 108–09. http://historical. library.cornell.edu/cgi-bin/cul.cdl/docviewer?did=cdl334&view=50&frames=0&seq=5

death, one of his early associates, Francis Jehl used the similarity between Edison's and Hertz's apparatus to claim Edison as a pioneer of wireless.[67] Thomson's biographer Woodbury made a similar claim for him.[68] In reality, neither Edison nor Thomson recognised the significance of what they observed at the time. In this, they were far from alone as many others before Hertz observed wireless phenomena without recognising it. These included such notables as American physicist Joseph Henry, British physicist Silvanus P Thompson and even Luigi Galvani, who, in 1780, noticed that a sparking electrostatic generator caused convulsions in a dead frog at some distance from it.[69,70]

Commenting on his abandonment of Etheric force, Edison observed, "If I had made use of my own work [on Etheric force in 1875] I should have had long-distance wireless telegraphy".[71] It must have been a galling realisation because Edison was constantly looking for such anomalous phenomena as sources of new inventions.[72] In other circumstances, Edison would have seized on the "curious result" in his 26 December experiment as the starting point and quite possibly, given his record, produced successful inventions from it, even if based on an erroneous theory. In this instance however, the 24 July 1876 rather than 26 December 1875 experiment signalled Edison's turning point, an end not a beginning.

The sequence of experiments from 22 November 1875 to 24 July 1876 trace Edison's 8 month path from success to failure. When he started on it in November 1875, he was just 28 years old, held 100 patents, was developing a public profile and in demand by people like Western Union who saw potential profit in his inventions. Less than 18 months later, Edison's prominence was such that he was invited to Washington to address the National Academy of Sciences at one meeting, members of the US Senate at another and meet US President Hayes.[73] In the midst of this success, Etheric force was a spectacular failure, one that might have ended his plans for the Menlo Park laboratory and certainly damaged Edison's credibility among scientists and sections of the wider community.

The Etheric force controversy divided scientists' opinion of him. He impressed some with his scientific ability, including George Barker and Joseph Henry, who had who made fundamental discoveries in electricity and magnetism and founded the Smithsonian Institution. They regarded Edison as a peer but others, remembering his Etheric force claims, refused to accept his work as legitimately scientific. One such

---

[67]Jehl, *Menlo Park Reminiscences*, 1, 89.

[68]D O Woodbury, *Elihu Thomson, Beloved Scientist 1853–1937*, 2nd ed. (1960).

[69]Charles Süsskind, "Observations of Electromagnetic Wave Radiation before Hertz," *Isis* 55, no. 179 (1964).

[70]Silvanus P Thompson (1851–1916) Thompson was a British physicist and writer who experimented with Etheric force in 1876 and concluded it was an induction effect. Thompson wrote biographies of Philip Reis and William Thomson (Lord Kelvin) and a widely used textbook on calculus.

[71]Dyer and Martin, *Edison, His Life and Inventions*. 578.

[72]Hughes, "Edison's Method."

[73]Baldwin, *Edison: Inventing the Century*, 98.

detractor was Henry Rowland, the first chair of physics at Johns Hopkins University and a vice president of the American Association for the Advancement of Science. In an influential and frequently quoted speech, Rowland referred to Edison (though not by name) as having a vulgar mind and to inventors as alien to the ideals of what he described as "pure science".[74] It was an influential sentiment that lasted decades. Not until 1927, near the end of his life (and still in the face of vigorous opposition) was Edison elected to membership of the National Academy of Sciences, the body that had honoured him in Washington in 1878.[75]

8.3 Thomas A. Edison, Inventor

Edison was unsuccessful in promoting his Etheric force theory because, when he moved from inventing to scientific theory making, he failed to use strategies he used successfully as an inventor. As the development of the carbon microphone showed, when Edison was working on an invention, failure was not something he avoided; it was something he actively pursued because, paradoxically, the success of techno-logical artefacts like the carbon microphone is a consequence of the thoroughness with which their creators pursue failure.

One way in which Edison used failure was through conceive-build-test sequences. These sequences form a cyclic process that usually began with a sketch that was built as a device to test. Failure of that device led to more sketches and more devices to test. A striking aspect of Edison's laboratory notebooks is the number of such ideas that he sketched. At peak output, Edison produced several hundred of these ketches each a month, a significant proportion of which were built, tested and their performance analysed.

When he turned to developing his Etheric force theory, Edison abandoned this strategy and so did not develop his Etheric force theory in the way he developed inventions. While some of the knowledge he acquired when inventing consisted of learning what worked and what did not, there was other, more valuable knowledge to be gained. When inventing, Edison identified aspects of the artefact that failed to work as intended and used this knowledge to produce more sketches to be built and tested, repeating the conceive-build-test cycle. Analysis of the resulting failures enabled him to identify those factors that were important to the success of an artefact and so determine criteria that a successful artefact should meet. When Edison started working on the carbon microphone, there were no existing microphones to provide models of performance so he needed to find what the relevant success criteria were. Since the transmission of articulate speech was his ultimate goal, Edison had to develop an understanding of human speech, inventing his own terminology in the

---

[74]David A Hounshell, "Edison and the Pure Science Ideal in 19th-Century America," *Science* 207, no. 4431 (1980).

[75]Israel, *Edison: A Life of Invention*, 468.

process such as calling sibilants "hissing sounds". This meant that in parallel with his artefact development process Edison also development a related success framework defining the success criteria that the artefact should satisfy.

Another, rarer, kind of knowledge that came from failures was the identification of anomalous phenomena he could exploit in new inventions. Sometimes there was an immediate connection as when Edison invented the Tasimeter, a device for sensing small changes in temperature, out of a failed microphone experiment.[76] At other times, he exploited failures that had occurred years before as he did when he applied the vibration sensitivity of carbon in his failed carbon rheostat to the carbon microphone.

Edison's pursuit of Etheric force began with observation of just such an anomalous phenomenon, the strange sparks he noticed on 22 November 1875. In contrast to his work on inventions like the carbon microphone, his subsequent laboratory notebook entries on Etheric force contain almost no conceive-build-test cycles. Hundreds of hours of testing separate the apparatus in Fig. 8.2 from that in Fig. 8.3 but the circuit for generating Etheric force sparks is identical. In contrast, Houston and Thomson improved on Edison's Etheric force apparatus by adding a Ruhmkorff coil, a type of high voltage transformer that gave them much larger Etheric force sparks.[77]

In his Etheric force experiments and theory development, Edison shifted from his invention strategy. Instead of trying many variations in the device under test, he applied many different tests to essentially the same device. Edison used these Etheric force experiments to explore Etheric force and build a systematic understanding of the phenomena. It was potentially useful knowledge but critically not directed towards supporting his public claim to have discovered a new force of nature.

8.4 Experiments in Science and Invention

Such a focus on exploration was not new to Edison and was to recur periodically when he encountered phenomena outside his existing knowledge. Steinle refers to such probing of phenomena as exploratory experiments.[78] The objective of exploratory experiments is to test many conditions to determine which of them affect the effect being studied, even though, at the time, there may be no adequate explanatory theory.

In such situations, Edison approached the new phenomenon by probing it to discover systems of regularities. Edison undertook a period of experimentation similar to that on Etheric force when seeking to understand more about electrical

---

[76]TAEB 3:1095A n2.

[77]Houston, "Phenomena of Induction."

[78]Friedrich Steinle, "Experiments in History and Philosophy of Science," *Perspectives on Science* 10, no. 4 (2002).

induction after his embarrassing failure of his automatic telegraph in England in 1873, due to induction effects.[79]

When no relevant scientific theories were available, Edison could use the systems of regularities developed through such exploratory experiments to provide the predictable results he needed to develop inventions. As he developed inventions, his experiments concentrated on improving the developing invention and testing it against increasingly arduous success criteria rather than expanding the range of success criteria tested. Edison directed these efforts towards the creation of an artefact, the invention, the success of which he judged against success criteria, often developed along with the artefact. An analogous experimental approach used by scientists and described by Steinle is the theory-driven experiment.[80] Like inventions, theory-driven experiments have a function: to support the related theory. Like inventions, the success of theory-drive experiments can be judged against success criteria, most significantly, eliminating alternative explanations by reducing the number of conditions required to produce the effect predicted.

A common outcome of theory-driven experiments is an artefact: the demonstration experiment, created to support the theory and eliminate alternative explanations. Unlike exploratory and theory-driven experiments, the outcome of the demonstration experiment should be known in advance with considerable certainty. Like theory-driven experiments, demonstration experiments are also judged against success criteria including the elimination of alternative explanations and predictability of results. Gooding describes Michael Faraday's development of an electromagnetic motor through many hours of refinement.[81] Shipped to various locations in Europe, the motor not only reliably reproduced the effect that Faraday described; it also supported his theory on the conversion of electricity into motion. In demonstration experiments like Faraday's, the epistemological objectives of science converge with the artefactual objectives of invention. For those who saw it in operation, Faraday's motor, a physical artefact, and the experimental procedure for using it, a non-physical artefact, convincingly supported both his claims about the phenomena and his theory.

Likewise, Edison used demonstrations convincingly to support his inventions. On New Year's Eve 1879, he demonstrated an artefact to the public, his Menlo Park laboratory complex lit by electricity.[82] In doing so, he was also, by implication, claiming that it demonstrated the validity of the knowledge that went into his electric lighting system. Put simply, it demonstrated that Edison knew how to produce light from electricity.

Both science and invention use an artefact, the demonstration, to support the validity of the underlying epistemological claims. Once there is a successful

---

[79]TAEB 2:321–336.

[80]Steinle, "Experiments in History and Philosophy of Science."

[81]Gooding, "Mapping Experiment as a Learning Process: How the First Electromagnetic Motor Was Invented.."

[82]Friedel, Israel, and Finn, *Edison's Electric Light: Biography of an Invention*, 119.

demonstration and particularly if others are able to replicate it, objections tend to focus on *what* knowledge it demonstrates rather than *whether* it has been demonstrated. Edison may have demonstrated electric lighting on New Year's Eve 1879 supporting his claim that he had the knowledge to produce light from electricity but the exact nature and extent of that knowledge did not go unchallenged. Two weeks after his demonstration, *New York Times*, still sceptical of Edison, carried an article which, while it did not dispute the New Year's Eve demonstration, questioned whether Edison yet knew how to produce lamps in sufficient number or economically enough to satisfy demand.[83] Similarly, because Edison could demonstrate, and others replicate, the novel qualities of Etheric force, the debate became one of what knowledge his demonstrations supported.

To be a successful artefact, Edison's incandescent lamp not only had to meet the success criterion of producing light from electricity, it also had to meet other success criteria including giving light for a time acceptable to purchasers, being produced in numbers sufficient to meet demand and selling at a price that was competitive with gaslight. Although Edison produced light from an incandescent lamp almost as soon as he started experimenting, it took well over a year to produce the first lamp that lasted 500 h.[84] As a precursor to identifying success criteria, Edison had to identify which factors were important and which to ignore. In fact, the key to the success of his incandescent lamp was Edison's realisation that the electrical resistance of the filament was critical. Three years earlier, his Etheric force theory failed, in part, because Edison ignored the need to identify which success criteria were critical to his theory and hence what counted as success or failure against those criteria.

8.5 Edison's Failure

In dealing with Etheric force, Edison failed both as an inventor and as a scientist. He failed as an inventor when he did not pursue the "curious result" of his 26 December 1875 experiment, missing the opportunity to develop long-distance wireless telegraphy. As a scientist, Edison failed to convince others of the validity of his theory. Instead, Houston and Thomson convinced him that their erroneous theory was correct. The key to understanding Edison's failure in both inventing and scientific theory making lies in the reasons why he accepted Houston and Thomson's theory and abandoned his own.

The simplest but least satisfactory explanation is that Houston and Thomson's superior theory persuaded Edison. The difficulty with this is that it is clear from Edison's notes that even part way through his July 1876 replication experiment, he did not accept their theory – and with reason. Contrary to Houston and Thomson's implication that Edison was ignorant of current electrical theory, Edison was an

---

[83]New York Times. "Thomas a Edison's Workshop: What a Visitor Saw and Was Told There – Some Discrepancies Noted." *New York Times*, 16 January 1880, 1.

[84]Friedel, Israel, and Finn, *Edison's Electric Light: Biography of an Invention*, 128.

expert on induction effects, had experimented extensively with them in 1873–74 and had exploited reverse currents, the basis of Houston and Thomson's theory, in inventions for automatic telegraphy.[85] It was not Houston and Thomson's reverse current theory that convinced Edison, it was their demonstration experiment.

An alternative, but still unsatisfactory explanation, implied by the rhetorical thrust of Houston's two papers is that Edison was merely a tinkerer in science, ignorant of current electrical theories. To accept this is to underestimate Edison. He may have had a limited formal education but he filled in the gaps in his knowledge by voraciously reading the current scientific literature and employing well-educated experts to give him personal tuition in areas in which he believed his knowledge was deficient. Further, Edison experimented night and day and had one of America's best-equipped electro-mechanical laboratories. In terms of effort, scientific knowledge, experimental expertise and facilities, he was in no way inferior to Houston and Thomson.

In his biography of Elihu Thomson, Carlson offers a social constructivist explanation of the Etheric force incident.[86] Carlson attributes Houston and Thomson's success to their exploitation of the unwritten rules of the scientific community and their use of an explanation that drew on conventional and widely accepted theories. Houston and Thomson, despite also having limited formal education, were also at pains to portray themselves as respectable scientific men in contrast to Edison who chose the path of populist self-promoter. As Pettit points out, the epistemological ascendency of science was still tenuous in this period, so American scientists sought to define their position, in part, by distancing their science from anything tainted with humbug or pseudoscience.[87] By discrediting the upstart Edison, Houston and Thomson enhanced the prestige of institutional science by distinguishing it from pseudoscience. In the process, they also enhanced their own prestige.

Carlson also argues that Houston and Thomson gained an advantage from publishing in the *Journal of the Franklin Institute*, rather than in newspapers, because of the scientific prestige the *Journal* brought. While this argument has some merit, it represents only part of the advantage Houston and Thomson gained from publication in the *Journal*. By writing their own paper, rather than relying on journalists, and publishing it in the *Journal*, they could exert far greater control over what it published, particularly given their prominent positions in the Franklin Institute. As an inventor, Edison sought to exert as much control as possible over the invention process and this was one reason for moving his laboratory to Menlo Park where he was freed himself from the sometimes-conflicting demands of the Newark manufacturing business in which he was a partner.

---

[85]TAEB 2:359, 2:361.

[86]Carlson, *Innovation as a Social Process: Elihu Thomson and the Rise of General Electric, 1870–1900*, 56–65.

[87]Michael Pettit, "The Joy in Believing: The Cardiff Giant, Commercial Deceptions, and Styles of Observation in Gilded Age America.," *Isis* 97, no. 4 (2006).

When it came to making his claims public, Edison chose breadth of audience over control of content and so had to rely on newspaper reporters and editors to convey his argument. His use of newspapers was motivated, in part, by self-promotion but it also served to attract potential investors and helped establish priority under US patent law. These strategies usually worked to Edison's advantage and in this instance, most accounts in the popular press were positive towards his claims. The risk in allowing others publish accounts of his experiments and theory was that they might be misrepresented or, as happened in the *New York Times* article, used to ridicule.

Houston and Thomson avoided these risks at the expense of a much smaller audience. By writing a paper and publishing it in the *Journal of the Franklin Institute,* they could carefully construct their arguments to achieve greatest impact. Their positions in the Franklin Institute also meant they could also prompt favourable editorial comment so that Houston's first paper was enhanced by an apparently independent editorial comment which asserted that "whatever there may be *remarkable* in the phenomena of so called Etheric Force, was described by Prof. Houston, [in 1871] previous to the *discovery* of Mr. Edison".[88] Despite these advantages, few read the *Journal of the Franklin Institute*, and critically, not Edison. Press clippings in Edison's papers suggest that he followed the progress of Etheric force debate, not through the *Journal of the Franklin Institute*, but in the popular press and *Scientific American*. The prestige of publication in the *Journal of the Franklin Institute* may have influenced some scientists as Carlson claims, but there is no evidence to suggest that it swayed Edison or caused him to abandon his Etheric force theory.

Houston and Thomson's use of the social structures of science and their efforts to discredit marginal science may have helped convince some scientists but they do not adequately account for Edison's change of mind. To understand what changed his mind we must look elsewhere.

8.6   Artefacts in Science

The course of the Etheric force debate shows that although their objectives are different, there are significant parallels between the process of inventing and the process of scientific theory making. For this reason, it is possible to treat one of the products of science, scientific theories, like the products of inventing, inventions. That is, we can treat both as artefacts. Like other kinds of artefacts, scientific theories are created for a purpose, can succeed or fail and can be judged against success criteria. Just as there are success criteria that can be applied to inventions, there are success criteria applicable to scientific theories: can the experiments that support theoretical claims be replicated; do the theories produce novel predictions; do they

---

[88] Houston, "Phenomena of Induction."

lead to new theories; and so on. If we can treat scientific theories as artefacts, it raises an interesting question. It makes no sense to ask of whether an artefact is true or not. The question for artefacts is, does it work. That is, does it meet its success criteria. If we applied this to a scientific theory, instead of asking about its truth we could be testing it against success criteria and, as noted above, getting the answer that success is situational so the theory may succeed (work) in some situations but perhaps not others. Newtonian physics is in widespread use because it meets the success criteria for many situations even if twentieth century developments have shown that it does not meet the success criteria for other situations.

Houston and Thomson convinced Edison to abandon his own theory and to accept theirs because they created better artefacts. One was a second paper constructed to resist alternative explanations. The other was their demonstration experiment that was stronger than Edison's because it was also designed to minimise alternative explanations.

Behind these artefacts lay another: Houston and Thomson's theory. Like their scientific papers and demonstration experiment, their theory was an artefact, intentionally created for a purpose. It began as a speculation into the cause of the anomalous sparks that Houston observed in 1871. In Houston's first paper, it is little more than an assertion but Houston did not leave it at this. With Thomson, he developed the theory, refining it to resist attack and alternative explanations. Just as better inventions use stronger materials, Houston and Thomson's built their theory on a stronger base both theoretically and rhetorically. Instead of Reichenbach's discredited Odic theory, Houston and Thomson built their theory on knowledge of electricity and induction already accepted by Edison and others working in the electrical field. Houston and Thomson's theory was more successful because it was more convincing and in this case, being convincing was a crucial criterion for its success as an artefact. Like inventions, Houston and Thomson developed these scientific artefacts over time by processes directed towards identifying success criteria and strengthening the artefacts to resist potential failure.

When inventing, past failures and the risk of future failures did not deter Edison. Writing about his work on electric lighting he observed, "Months of intense watching, study and labour are required before commercial success – or failure – is certainly reached."[89] When he turned to developing his Etheric force theory, Edison did the reverse and instead acted as though his initial intuition was enough. He skipped the "months of intense watching, study and labour" when developing his theory, publishing it a few days after his initial observations. Like his Etheric force experimental apparatus, which also did not develop, Edison's theory went no further than the assertions he made at the end of November 1875. Even those few parts of his theory that he did develop in the following weeks went unpublished, so public knowledge of his theory remained as he developed it in the first few days after his 22 November discovery. Edison was fond of announcing, often on the slightest of

---

[89]TAEB 4:1570.

evidence, that he was "going to" invent something. Despite this, it was virtually unthinkable that he would have demonstrated an invention in public before spending many months of development, probing for potential failures and strengthening it to resist them, yet this is what he did when he publically announced his Etheric force theory after only a few days of development.

In approaching the subject through exploratory experiments, Edison was following the strategy of his experimental hero, Michael Faraday, (who Edison referred to as "the Master Experimenter"), and the theoretical source of the Etheric force concept, Karl Reichenbach.[90] During the 1840s, a resurgence of interest in Mesmer's animal magnetism led both Faraday and Reichenbach to experiment on the effects of magnets on nominally non-magnetic materials (diamagnetism). Despite the similarity of subject, Faraday and Reichenbach's theoretical paths diverged radically.

In addition to examining the effects of magnets, Reichenbach, extended his experimental work to other substances including crystals. Typical of these is Reichenbach's description of a battery of exploratory experiments to establish the effect of his new force (Odic force) in chemical reactions.[91] He began by noting that a wire dipped into a solution of sodium bicarbonate felt warm. Since, as he acknowledged, it might be objected that electricity generated in the process could have produced the effect, he explored alternatives. This exploration included using people Reichenbach called "sensitives" because, unlike normal people (including Reichenbach), they claimed to be able to detect or see emanations associated with Odic force. One of these sensitives, Mlle Reichel, claimed to observe light around the wire in a darkened room. As Edison did later with Etheric force, Reichenbach rapidly moved from exploratory experiments to a broad theoretical claim about Odic force, concluding that "Everywhere, therefore, even where mere solution of water or combination of water of crystallisation occurred, chemical action developed, in an active state, the new force".[92]

Faraday's approach, in contrast, was much more cautious, even though he was working with the similarly surprising phenomenon of diamagnetism. In November 1845, Faraday built a massive horseshoe electromagnet from a large chain link (95 mm diameter, 1170 mm axial length and weighing 108 kg).[93] The magnetic field it produced was so great that Faraday noted that he could not pull a piece of iron off the poles when energised. Using this electromagnet, Faraday conducted a series of experiments at the end of 1845 on magnetic (mostly ferrous) materials and a wide

---

[90]Dyer and Martin, *Edison, His Life and Inventions.* 101.

[91]Karl Reichenbach, *Physico-Physiological Researches on the Dynamides or Imponderables (Magnetism, Electricity, Heat, Light, Crystallisation, and Chemical Attraction) in Their Relation to the Vital Force*, trans. William Gregory (London: Taylor, Walton, & Maberly, 1850), 113–20.

[92]Ibid., 116.

[93]Michael Faraday, *Faraday's Diary: Being the Various Philosophical Notes of Experimental Investigation Made by Michael Faraday During the Years 1820–1862 and Bequeathed by Him to the Royal Institution of Great Britain. Now, by Order of the Managers Printed and Published for the First Time under the Editorial Supervision of Thomas Martin M Sc*, ed. Thomas Martin, 7 vols., vol. IV (London: G. Bell and Sons Ltd., 1933), 310.

range of diamagnetic (nominally "non-magnetic") materials including glass, gases, chemical compounds, natural materials and non-ferrous metals. Specimens were suspended between the poles of his massive magnet either directly or in a glass tube and their movement observed when power, and hence magnetism, was applied.[94] Faraday lists many that rotated under the applied magnetic field, diamagnetic effects occurring because, when subject to an external magnetic field, conduction in the substances produces an opposing magnetic field.

Although Faraday conducted many such experiments, his theoretical claim was quite modest, no more than the observation that all the substances he tested exhibited magnetic effects. He did, however offer cautious speculation on the phenomenon. After observing the diamagnetic reaction of organic materials to his powerful magnet, he commented, "If a man could be in a Magnetic field, like Mahomet's coffin, he would turn until across the magnetic line, provided he was not magnetic".[95]

What distinguished Edison and Reichenbach from Faraday was not the way in which they conducted their research, which was similar, but extent of the theoretical claims they were prepared to make based on exploratory experiments alone.

In confining himself to an initial theoretical insight and exploratory experiments, Edison succumbed to the cultural allure of science. In public, the work of the scientist is often portrayed as heroic exploration with flashes of insight. In the privacy of their laboratories, scientists do something different, something that is systematic, unspectacular and, at times, tedious. Such systematic, unspectacular and tedious research was also what Edison, the inventor, spent much of his effort on. Published scientific accounts, for the most part, omit descriptions of failures and the experimenters involved even find it hard to recall why past difficulties were even difficulties. The absence of published accounts and limited recall highlight the value of following the processes scientists and inventors use as recorded in their laboratory notebooks. These contain records of what the scientist or inventor observed and believed at the time an experiment was done. Most importantly, they are untainted by knowledge of subsequent events, revealing the erroneous theories, failures and blind alleys that tend to be omitted from retrospective accounts. Indeed, part of the persuasiveness of published scientific papers lies in the way that they omit the failures and shortcomings overcome.

The demonstration experiment in Houston and Thomson's second paper represents the end of a development process, a refined artefact, strengthened to resist attack. As he replicated it in July 1876, Edison could not know what had been done to create it, nor what failures had been overcome, but to replicate it he had to engage with Houston and Thomson's delicate experimental apparatus. In order to master the experiment, Edison needed, to a certain extent, to think like Houston and Thomson.

Arriving in triumph at the 1891 Chicago World's Fair, Edison drew an analogy between science and invention. In the late nineteenth century era of heroic

---

[94]Ibid., 310–95.

[95]Ibid., 325–26.

exploration and colonial expansion, Edison chose to do this by comparing scientists, inventors, geographers and explorers. He was quoted as saying,

> There is as much difference between an inventor and a scientist as there is between an explorer and a geographer ... Of course scientists may be inventors and inventors may be scientists. And explorers may write geographies, but they seldom do. The inventor discovers things and then the scientist steps in and tells or tries to tell what it is that has been discovered.[96]

By this time Edison had come to identify his work as an inventor with that of explorers. It was a somewhat myopic view. To be an inventor is to create artefacts and geographers, like inventors, also create artefacts such as maps. Edison's success as an inventor may have been due acting as explorer, coming to initial insights and working through exploratory experiments but, as he acknowledged himself, success at inventing required more than an initial flash of inspiration. It required much labour and in this, Edison the inventor acted like a geographer. The view of science Edison expressed in Chicago is erroneous, not because it was fundamentally wrong, but because it is too narrow.

A consequence of this narrow view, was that the conceive-build-test sequences characteristic of Edison's approach to inventing were absent from his Etheric force experiments. Edison also failed to follow the inventor's approach of using failure to seek the weaknesses and vulnerabilities of his theory. Had he done so, the knowledge he could have used it to identify relevant success criteria and thus direct his efforts towards strengthening and refining his theory.

Crucially, although he seems to have convinced himself that Etheric force sparks were not due to induction, he did not develop his theory or a demonstration experiment to counter the induction argument expressed by so many of his opponents and the view that Edison himself had initially held. Despite this opposition, he offered no better demonstration to counter the induction argument than the repetition of a few tests he devised in the first few days after observing the Etheric force phenomenon.[97]

As an artefact, his demonstration experiment was weak and vulnerable to the failure that it experienced, yet Edison was more than capable of developing a demonstration experiment to show that Etheric force was not induction. Instead of directing his efforts to countering alternative explanations, his demonstrations concentrated on the more spectacular aspects of Etheric force. These may have amazed his audience but they failed to meet the success criteria for a successful scientific artefact. To be successful in this respect, Edison's demonstrations needed to convince knowledgeable observers like the inventor, John E Smith and physicist, George Barker, both of whom remained sceptical.

Despite this limitation, Edison's demonstrations of the spectacular aspects of Etheric force were artefacts with other functions, notably to publicise Edison. In this, they succeeded. Reports of Edison's spectacular demonstrations may have done little to enhance Edison's reputation among scientists but they contributed to building his public image as a creator of remarkable things.

---

[96]Chicago Daily Globe. "Arrival of Thomas A. Edison." *Chicago Daily Globe*, 13 May 1891.
[97]TAED NE1691:18.

Edison's demonstrations supported his claims about the phenomena but not his theory. Because Edison produced no effective demonstration supporting his theory, Houston and Thomson's demonstration not only supported theirs, it also became a de facto success criterion. In the absence of a counter demonstration from Edison, theirs filled the vacuum and so was even more convincing.

While Houston and Thomson succeeded in convincing Edison, their approach, like Edison's, had a negative side. The negative for Edison was to emphasise exploratory experiments and neglect theory development. Houston and Thomson did the opposite, neglecting exploratory experiments in favour of developing their theory and a strong demonstration experiment. Houston and Thomson accepted, apparently without question, their anomalous observation that non-metallic humans affected what they believed to be electrical induction. They also did not explore the more spectacular aspects of Etheric force that Edison excelled in demonstrating. Forty-five years later, a fellow teacher at the Philadelphia Central High School, described the 22-year-old Elihu Thomson excitedly running through their school building and onto the roof, drawing sparks from all manner of metal objects.[98] This was a remarkable observation yet Houston and Thomson seem to have ignored it. Prior to this, induction effects, the basis of their theory, had been observed only over distances of much less than a metre. In Snyder's description, induction appears to be causing sparks over distances of many meters and through several floors of the building. The result of not exploring these anomalous phenomena was that Houston and Thomson, like Edison, missed the opportunity to pioneer wireless communication for Thomson's sparks were evidence that they were transmitting something considerable distances without wires.

8.7 Science and Technology

The battle over Etheric force was fought out, not by scientists working within Henry Rowland's concept of pure science but between inventor-scientists, Edison on one side and Houston and Thomson on the other. Both Carlson and Israel attribute Edison's failure in the Etheric force debate to his scientific naivety, as evidenced by his use of popular newspapers, rather than scientific journals to publish his claims.[99,100] The details of the dispute suggest that, while Edison's naivety in this respect was a factor, his failure had more to do with his misunderstanding of the processes needed to have a scientific theory accepted, processes that bear a close relationship to those that he used to create successful inventions.

---

[98]Monroe B Snyder, "Professor Elihu Thomson's Early Experimental Discovery of the Maxwell Electro-Magnetic Waves," *General Electric Review* XXIII, no. 3 (1920).

[99]Carlson, *Innovation as a Social Process: Elihu Thomson and the Rise of General Electric, 1870–1900*, 63.

[100]Israel, *Edison: A Life of Invention*, 114.

The Etheric force controversy also demonstrates the importance of artefacts in convincing others to accept a scientific theory. Edison's artefacts were weak and did not convince, while Houston and Thomson's were stronger and, crucially, convinced Edison. Edison failed to advance his claim to have discovered a new force of nature, not so much because of defects in his theory, but because he succumbed to a concept of science, a cultural myth, that portrays science as fundamentally different from inventing. In so doing, he acted as though science, unlike inventing, did not involve building artefacts, perhaps because the key artefacts of science, theories, were not like the physical artefacts he invented. If this is so, it is surprising since Edison also created other successful non-physical artefacts notably patents to protect his inventions. Like Houston and Thomson's scientific papers, Edison's patent applications were carefully constructed to resist rivals.

Edison's reliance on exploratory experiments in relation to Etheric force suggests that in 1875 he believed that scientists were explorers, not geographers. By the time he arrived in Chicago in 1891, he had come to believe that it was inventing, not science, that required an explorer's drive and skill. Despite thinking that explorers (i.e. inventors) were rarely geographers (i.e. scientists), had he looked at the history of science he would have noticed that many prominent scientists were also inventors including Galileo, Newton, Kelvin and his own hero, Faraday. That they were successful at both should not have surprised Edison if he recognised that they used the same processes to create physical artefacts, like inventions, as they did when creating the artefacts of science including papers, demonstrations and theories.

A further aspect of Edison's work on Etheric force and Reichenbach's research into Odic force is that both focused on exploring the properties of their discovery rather than testing and developing their theories. One possible reason for this is that both Edison and Reichenbach were successful inventors. Reichenbach built his wealth from inventions, notably in iron smelting, initially embarking on scientific research associated with his inventions and commercial interests. Edison and Reichenbach both rapidly conceived, and accepted, a theoretical explanation (Etheric force and Odic force) for the phenomena they observed, then concentrated on researching its properties and effects.

As inventors, they were primarily concerned with exploiting the phenomena they observed rather than building stronger theories to convince others. The inventor's search for knowledge parallels that of the scientist, with a significant exception. Both are theory-building exercises but for the inventor, having an erroneous theory may not be fatal to the invention since the use of failure as a tool enables inventors to overcome the consequences of erroneous theories. The theory might be erroneous but the test of a physical artefact is whether the invention fulfils the success criteria in its success framework, not the truth of the theory used to create it. The success of the invention depends on whether the effects of the phenomena exploited in the invention are repeated consistently, not that the theoretical explanation used by the inventor is correct.

The criteria for the success of scientific theories are not the same as those for inventions.

Part III
Edison's World

Chapter 9
Thomas Edison and Patents

9.1 Patents Make Edison

Without patents and the patent system, Thomas Edison could not have become a professional inventor, nor without patents, could he have become an inventor-entrepreneur. Understanding Edison as inventor and entrepreneur requires understanding patents and the patent system.

The most striking thing about Edison's patents is their number, Edison being issued 1084 utility patents in the United States, more than any other inventor.[1] He applied for his first patent in 1868 at the age of 21 and his last in 1931, the year that he died. Not only did his inventive career span 63 years but his patent output was prodigious, averaging a patent every 21 days over that time and peaking one every 3.4 days in 1882 (Fig. 9.1). In addition to these successful patent applications, Edison also submitted a further 500–600 unsuccessful applications that were either rejected by the United States Patent Office or withdrawn by him.[2]

The number of Edison's patents and the rate at which he produced them are both impressive but so is wide range of subjects addressed by his patents. While he is best known for his inventions relating electric lighting and sound recording, his inventions cover much more. Edison's inventions are so numerous and varied that it is difficult to summarise them, but his lesser known but significant inventions include

---

[1]Edison was also issued with nine design patents (called registered designs in some countries). In the United States, design patents register the ornamental (non-functional) design of a functional item. Unlike utility patents, the design patented must have no practical utility. In general, when we speak of patents in relation to inventions we are referring to utility patents so I refer to utility patents simply as patents.

[2]Thomas A. Edison Papers. 2019. "Edison's Patents." [web page]. The Thomas Edison Papers, Rutgers, The State University of New Jersey. http://edison.rutgers.edu/patents.htm

© Springer Nature Switzerland AG 2019
I. Wills, *Thomas Edison: Success and Innovation through Failure*, Studies in History and Philosophy of Science 52, https://doi.org/10.1007/978-3-030-29940-8_9

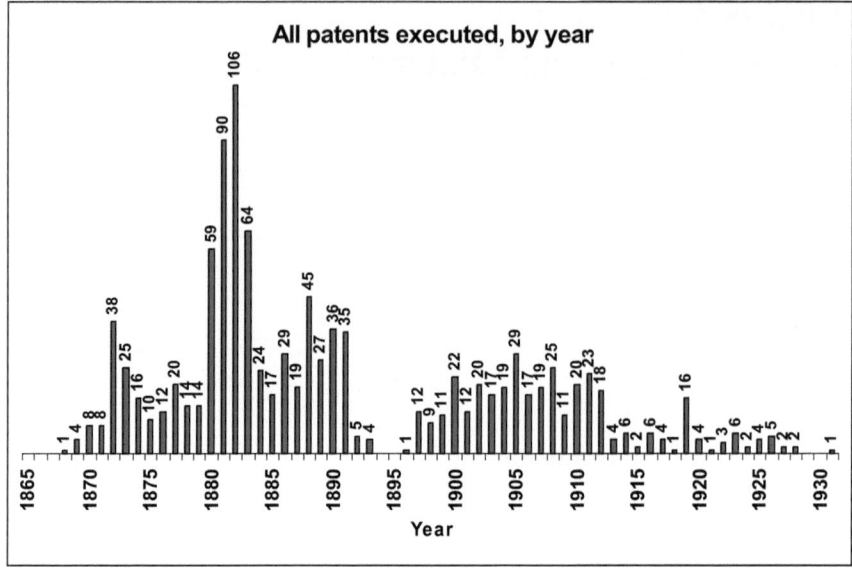

Fig. 9.1 Edison's patents by year of execution. (Compiled from ibid)

improvements to the stock ticker, alkaline storage batteries, large scale cement and concrete manufacture, and the electric pen.[3] Table 9.1 illustrates the remarkable breadth of his inventive output.

9.2 The Patent System

Patents are not inventions. Patents are a legal concept granting a temporary monopoly to an inventor for a fixed period, (17 years in the United States during Edison's career). This monopoly right can be, and usually is, assigned to others such as corporations who then use the monopoly to exploit the patented invention commercially. The term patent is derived from "letters patent" referring to the government document that makes public (that is, patent) the details of the patentee's invention and their claims about it. Governments intend that the publication of patents will encourage innovation and in return for this public knowledge, grant the inventor a

---

[3]The electric pen was the basis for the Mimeograph and later the electric tattoo pen. Copies of documents were made with the electric pen by writing or drawing on a specially treated paper master sheet, the vibrating point of the pen perforating it. Multiple copies could then be made (Edison claimed 5000 from one master) in a special press which forced ink through the perforations. Initially Edison and his partner, the machinist Joseph Murray, manufactured and sold the pens and duplicating presses themselves but Edison later sold the patent rights to A B Dick who renamed it the Mimeograph and developed it into a substantial business.

Table 9.1 Edison's utility patents grouped by subject and ordered by the date of execution of his first patent application

Patent subject	Number of patents	% total	Year of execution of first patent
Voting The Electrographic Vote Recorder, Edison's first patent	1	0.1%	1868
Telegraph Includes the quadruplex, telegraph relays, automatic telegraph, printing telegraph, (stock ticker) and acoustic telegraph	141	13.0%	1869
Batteries Includes primary and secondary battery designs and methods of manufacturing alkaline cells	107	9.9%	1872
Electric pen Patents that formed the basis for the mimeograph	6	0.6%	1876
Phonograph Includes the phonograph, methods of producing recordings, and improvement to phonograph components	178	16.4%	1877
Telephone Includes the carbon microphone, electromotograph receiver	39	3.6%	1877
Electric light and power Includes incandescent and fluorescent lamps, generators, motors, cables, electric meters and lightning (surge) protection	388	35.8%	1878
Ore extraction and mining Includes magnetic ore extraction, crushing and milling machinery, briquette manufacture and improvements to enable machines to work in dusty atmospheres	80	7.4%	1880
Transportation Includes electric railways, electric cars, railway signalling and a flying machine	38	3.5%	1880
Electroplating Includes methods of electroplating and applications such as the production of finely divided metals and thin metal sheets	30	2.8%	1882
Manufacturing processes Includes the manufacture of glass, wire and production of plastic from vegetable fibre	7	0.6%	1882
Mechanical power transmission Includes shafts, belts and gearing	9	0.8%	1882
Wireless communication Includes telegraphy by induction and the antenna patent sold to Marconi	6	0.6%	1885

(continued)

Table 9.1 (continued)

Patent subject	Number of patents	% total	Year of execution of first patent
Motion pictures Includes a motion picture camera, projection screen and producing coloured images	9	0.8%	1891
Portland cement and concrete Includes production of Portland cement, cement kilns and applications of concrete	32	3.0%	1900
Military Three projectile designs	3	0.3%	1916
Moulding processes Patents on methods for mass production of moulded articles other than in connection with the phonograph	7	0.6%	1919
Protection of ferrous metals Zinc rich paint for protecting ferrous metals	1	0.1%	1919
Rubber production Two patents dealing with synthetic and chlorinated rubber	2	0.2%	1923
Total utility patents	**1084**	**100%**	

Compiled from Thomas A. Edison Papers, "Edison's Patents"

fixed term monopoly on its exploitation. In practice, the patent monopoly can also have the reverse effect, inhibiting innovation by other inventors in related areas.

As well as encouraging innovation, during the eighteenth and nineteenth centuries the American government intended that publication of patents would educate other inventors who, rather than merely copying the inventions, would emulate them, that is, develop and improve of patented inventions.[4]

To encourage emulation, inventors were required to submit a model of their invention with their patent application, models helping assessment of patent applications and being publicly displayed if a patents was issued. The models could then be studied by other inventors who, it was intended, would be encouraged to emulate the inventions. As inventions became more complex and non-mechanical, particularly those employing electrical and chemical concepts, the relevance of mechanical models diminished so the requirement for patent models was abandoned. With this, inventors turned to patent gazettes and magazines such as *Scientific American* for details of new patents, both to emulate and to identify potential threats to their existing patents. Until 1948, *Scientific American* was published by Munn & Company, one of America's largest patent agencies.

All patents may be created equal but they are not all equal in their effect. Most patents are issued for relatively minor improvements to existing inventions. These patents may be valuable but one group of patents, which I refer to as enabling

---

[4]Brooke Hindle, *Emulation and Invention* (New York: New York University Press, 1981).

patents, are historically significant because they make an invention or technology feasible. The rightful inventor of the telephone may be open to dispute but it was Alexander Graham Bell's first telephone patent that enabled telephone technology to develop.[5] Similarly, Edison's high resistance incandescent lamp patent made reticulation of electrical power from a central station, and hence electrical power utilities, feasible.[6] While many others, including Joseph Swan in England, produced incandescent electric lamps before Edison, none of their inventions had the same enabling effect on the electric power industry. (Swan, a Newcastle pharmacist, devised several incandescent lamp designs, demonstrating one in 1860 and receiving a British patent in 1878, the year before Edison's US patent.)

Enabling patents may also be controlling patents. By enabling the development of a new technology, the monopoly conferred by enabling patents often gives its inventor (or assignee) control of the technology, at least initially. Edison's patents for the Phonograph and incandescent lamp gave him initial commercial control of both these technologies but in both cases the control was relatively short lived, being diminished by later patents. These enabled derivative, but technically superior, technologies to be introduced; Berliner's disc gramophone in the case of the Phonograph and Tesla's polyphase alternating current electrical system in the case of electric lighting.[7,8,9,10]

Patents are a common legal means of creating a monopoly for an invention but they are not the only way of doing this. In the 1840s, Henry Bessemer invented a method for making cheap imitation gold (gilt) paint.[11] Bessemer is most famous for invention of the Bessemer process for the conversion of iron to steel but when competitors flouted his patent, he determined to use a non-patent approach to protecting his gilt paint invention. Instead of patenting it and making the process public, he constructed a purpose-built factory in which the paint was made in separate stages in different parts of the building, the secrets of whole process being known only to Bessemer and members of his family. By using secrecy rather than a patent, Bessemer was able to maintain a monopoly on gilt paint for about 35 years, twice that available through patent protection and without the risk and cost of patent infringement.

When, in the 1890s, Edison became frustrated with the American patent system, he contemplated using secrecy to protect his inventions but abandoned the idea because of its limited usefulness for the kinds of inventions he was producing.[12]

---

[5]Bell. Improvement in Telegraphy.

[6]Edison. Electric Lamp.

[7]Phonograph or Speaking Machine.

[8]Electric Lamp.

[9]Emile Berliner. Gramophone. US Patent 372,786, filed 4 May 1887, and issued 8 November 1887.

[10]Nikola Tesla. System of Electrical Distribution. US Patent 381,970, filed 23 December 1887, and issued 1 May 1888.

[11]Henry Bessemer (1813–1898) British inventor. His other inventions included sugar processing.

[12]Israel, *Edison: A Life of Invention*, 318.

Secrecy is most effective for protecting manufacturing processes, like Bessemer's gilt paint process, but ineffective for protecting artefacts that are sold. Secrecy would have provided no protection for Edison's tinfoil Phonograph. Secrecy also complicates licencing inventions to others. Finally, as Edison admitted, "Everybody steals in commerce and industry. I've stolen a lot myself."[13] Achieving a monopoly through secrecy required an exceptional invention and taking exceptional measures, as Bessemer had done.

9.3   Patent Priority

There are three significant dates related to patents: the date of execution, the date of application and the date of issue. The date of application is the date on which the patent office received the inventor's application for a patent. The date of issue is the date on which the patent is issued. In Edison's case, the period between date of application and date of issue varied widely, the shortest being 13 days and the longest almost 18 years. Typically, issue of a patent took 12–18 months.

The third significant date is the date of execution, that is, the date on which the inventor signed the patent application. This date is more representative of the invention processes because it is closer to the date of invention than date of application. Although the date of application for Edison's patents was generally a few days of execution (a median of 9 days), in some cases the delay was several months.[14]

Patents are issued to the first creators of novel artefacts. Simple as this may appear, in practice the criteria for deciding who was first varied from country to country in the nineteenth century. An aspect of the American patent system that influenced Edison greatly was that patents were issued to those who could prove they were the first-to-invent. In this, American patent law differed from that of most other countries where the criterion was the first person to file a patent application. To assist in establishing priority, the American system allowed for the filing of a document known as a caveat before filing of a patent application. The caveat was a notification to the Patent Office of the inventor's intention to file a patent application within a year and was held in a confidential file where it could provide formal evidence of priority if needed.

A caveat established priority formally but there were also informal ways of doing this. In Newark and later at Menlo Park, Edison did not work in secret and allowed

---

[13]Rosanoff, "Edison in His Laboratory."

[14]As United States patents show the date of application and issue, not the date of execution, the information for date of execution of Edison's patents is taken information published by the Edison Papers project (Thomas A. Edison Papers, "Edison's Patents".)

outsiders visit his laboratory freely, even demonstrating his latest creations. In fact, he said of his laboratory, "It was kind of a public place".[15] Doing this might appear reckless, since visitors might steal his ideas, but Edison's visitors were also potential independent witnesses to support future priority claims. Edison also preferred to announce new developments in the popular press, also helping to establish priority and useful for attracting potential investors.

The fact that the United States used the first-to-invent criterion while other countries used first-to-file could lead to problems. In 1877, as he was working on the Phonograph and before he had filed a patent application or caveat, Edison allowed his associate, Edward Johnson, to describe his invention in several newspapers. Then, before he filed his patent application, he demonstrated the Phonograph in the Scientific American offices, *Scientific American* publishing a detailed description of the Phonograph with drawings in its next issue, published 2 days before his patent application.[16,17] The publication of these reports established Edison's priority as first-to-invent the Phonograph in the United States, but in other countries one of the tests applied to a patent application was whether, prior to the patent application, details were known that would permit a person with suitable skills to create the invention being patented. United States patent law still uses this as one of its criteria (referred to as prior art), stating that an inventions cannot be patented if "the invention was patented or described in a printed publication in this or a foreign country".[18] Making details public, including applying for a patent in another country, could be used to argue for the existence of prior art, making that the invention not patentable in the second country. Edison was faced with this argument in other countries when he tried to obtain patents on the Phonograph, the Johnson and *Scientific American* articles being used as evidence against Edison.[19]

In the United States, patents recognise and name only individuals as inventors, not organisations, even if the inventors are employees of an organisation and the organisation owns the patent through a contract of employment. Most, but not all, of Edison's patents name him as sole inventor. The naming of individual inventors in patents had a significant influence on the way in which Edison worked and on his relationship with employees. If a patent named only Edison but it could be shown that others were joint inventors with him, it might be declared invalid. To reduce this risk, Edison's policy was that his ideas alone were embodied in his patents, while his employees, like Charles Batchelor, worked as experimental associates or, like John Kruesi, as builders of experimental models. There are several accounts, including one from Tesla, of Edison refusing to consider suggestions from his employees.

---

[15]TAED TI1:22.

[16]Edison. Phonograph or Speaking Machine.

[17]Scientific American, "The Talking Phonograph."

[18]US Department of Commerce, *Manual of Patent Examining Procedure*. clause 706.02.

[19]TAEB 3:1150n2.

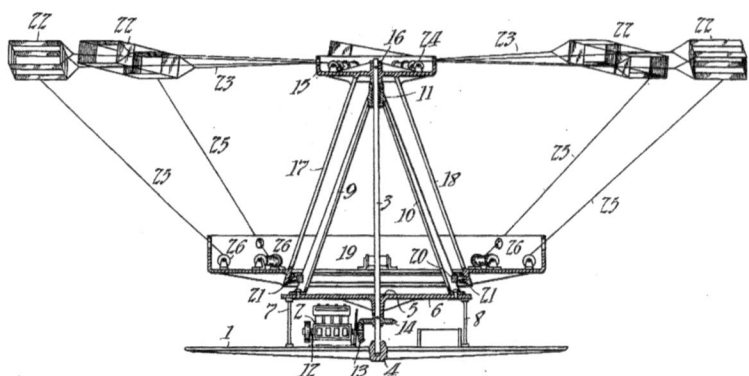

Fig. 9.2 Edison's flying machine. Fundamental problems with actions and reactions mean that it could never have flown even if the box kite arrangement had provided enough lift, which is also unlikely. (Thomas A Edison. Flying Machine. US Patent 491,993, filed 16 November 1908, and issued 20 September 1910)

While this might be interpreted as Edison wanting to deny that others might have valuable ideas to contribute, had he made use of such advice in a patentable invention he would also have needed to name the other person as joint inventor. Instead, Edison seems to have decided for the most part to make use of his associates' skills but not their ideas.

To be patentable, an invention must be novel, useful and not obvious. Under current United States law, novelty is defined in the negative rather than positive terms, so, for example, an invention is novel if it is not known or used by others before the application, not obvious and has not been described in a foreign patent or publication.[20] While such a negative definition may be a convenient way to administer patent law, it is awkward historiographically. In Chap. 4, I propose the use of functional systems as a means for identifying novelty that overcomes this problem.

Despite this limitation, patents provide a relatively objective way of judging novelty primarily because patent issuing authorities do not enquire into whether or not the invention submitted achieves the functions claimed for it, only whether or not it is novel. Edison conceived many more novel inventions than he successfully patented. Some, like the condenser microphone conceived as he was developing the carbon microphone, were potentially viable, while others, like his flying machine (Fig. 9.2), were not. Even omitting these from the present analysis and confining it to issued patents, allows insights into Edison's approach to innovation evident in patterns in his patents.

---

[20]US Department of Commerce, *Manual of Patent Examining Procedure*. Appendix L 35 U.S.C. 102.

9.4   Why So Many Patents?

Edison's 1084 American patents may be the greatest number for a single American inventor but he worked in an age that saw other inventors amass similarly impressive numbers. Elihu Thomson, Edison's critic in the Etheric force dispute, was issued with 696 patents while Elmer Sperry's total ran to over 350. They were far from isolated cases to the extent that it has been claimed that, "No other nation has displayed such inventive power and produced such brilliantly novel inventors as the United States during the half century beginning around 1870. [It was a] gigantic tidal wave of human ingenuity."[21]

Impressive as the numbers of patents might seem, they do not represent an equal number of viable inventions. In 1869, the United States Commissioner of Patents estimated that only 10% of patents he issued had commercial value.[22] Examination of Edison's 1084 patents suggests that many of them also had little chance of commercial success, raising the question as to why Edison and other inventors would go to the effort and significant cost of applying for patents on inventions that were commercially or technically questionable.

One reason is that many successful inventions are not created as a single, finished concept, but emerge through a development process, stages in which are marked by patents, with later patents addressing failures in earlier patents. Edison's crucial incandescent lamp patent was not his first for such a lamp but his tenth.[23]

A second motivation is that acquiring patents can be strategic commercially, a form of defence, even when the patented invention is not developed beyond the patent stage. This appears to be the case with some of Edison's microphone patents.[24] With these, Edison's objective appears to have been to defend his chosen course by covering as many alternative solutions as possible, without necessarily intending to develop them into viable commercial devices. Possessing these patents inhibited potential competitors from developing the principles patented for the duration of the patent. Since Western Union had a near monopoly over telegraphy at the time, also acquiring patents for alternative technologies meant that only the technologies Western Union favoured could be used in in the telegraph industry. A potential competitor to Western Union was therefore compelled to use Western Union technology, further strengthening Western Union's monopoly position.

A final reason for patenting questionable inventions is that the inventor gained personal kudos from accumulating many patents. For the independent inventor-entrepreneurs of this period such as Edison, Thomson and Sperry, more patents

---

[21]Hughes, *American Genesis*, 13–52.

[22]Basalla, *The Evolution of Technology*, 69.

[23]Edison. Electric Lamp.

[24]Jehl, *Menlo Park Reminiscences*, 1, 143–54.

not only led to greater financial rewards but also increased their public reputation and self-esteem. These inventors "tended to count patents as well as – perhaps more than – money as symbols of success".[25]

9.5   Edison's Patents

Table 9.2 lists Edison's most significant inventions in order of the date of first patent application. Although these are significant inventions they were not all commercially successful: Edison's first patented invention, the vote recorder, was rejected by the legislators who where its intended users; the tinfoil Phonograph was immensely successful at raising Edison's public image but had essentially no commercial success until he developed the wax cylinder Phonograph 10 years later. The loud-speaking telephone succeeded in avoiding Alexander Graham Bell's patents but was only a marginal success commercially. Conversely, Edison's iron ore inventions were successful technically but a commercial disaster for Edison after cheaper high-grade ore became available. Finally, the concepts behind Edison's fuel cell patents were innovative, especially for the era, but were technically and commercially unsuccessful.

For Edison, the principal success criterion for a patented invention was that it was a success commercially. If we apply this criterion to Table 9.2, four of the inventions (the vote recorder, the tinfoil Phonograph, fuel cell and magnetic ore milling) were not commercially successful, while the loud-speaking telephone and disc Phono-graph were of only marginally successful. Omitting the unsuccessful and marginal inventions leaves just 12 on which Edison's success and fame as an inventor rests.

While some of these 12 inventions are associated with many patents (96 in the case of wax cylinder Phonograph), each invention, as a radical innovation, is represented by one or perhaps a handful of patents. The remaining patents, over 1000 of them, can be divided into broad categories:

1. Minor innovations (improvements) to earlier inventions.
2. Developmental stages of later inventions.
3. Defensive patents.
4. Fanciful inventions.
5. Commercially unsuccessful inventions.
6. Inventions made successful by others.

The largest group consists of minor innovations (improvements) to existing technologies. The size of this group can be judged from his electric lighting patents. Over his lifetime, Edison was issued with 388 patents relating to electric lighting, of which 165 (42%) were for one invention, the incandescent lamp and aspects of its manufacture.

---

[25]Hughes, *American Genesis*, 24.

Table 9.2 Edison's inventions and related patent statistics

Invention	Year of first patent application	Number of patents relating to the invention	Number of patents in first 10 years	Number of patents in the first 17 years (duration of first patent)
Vote recorder	1868	1	1	1
Automatic telegraphy	1869	45	45	45
Stock ticker	1871	30	30	29
Quadruplex telegraph	1874	4	3	4
Electric pen	1876	6	6	6
Telephone transmitter	1877	27	26	26
Tinfoil phonograph	1877	2	2	2
Electric lamp	1878	165	155	162
Loud-speaking telephone	1879	1	1	1
Electric generator	1879	100	93	99
Electric light and power system	1880	54	54	54
Ore milling and processing	1880	80	10	30
Fuel cell	1882	3	3	3
Wax cylinder phonograph	1887	96	71	90
Motion pictures	1893	9	3	4
Cement	1899	32	6	29
Storage battery (alkaline)	1900	107	73	88
Disc phonograph	1910	18	13	18

The list of significant inventions comes from Thomas A. Edison Papers. 2019. "Inventions." [web page]. The Thomas Edison Papers, Rutgers, The State University of New Jersey. http://edison.rutgers.edu/inventions.htm. The number and dates of patents is derived from "Edison's Patents"

The second group of patents represents developmental stages on the way to successful inventions. For example, Edison applied for nine patents for incandescent lamps before the controlling patent in which identified high filament resistance as the critical success criterion for electric lighting.[26] Edison's controlling patent may have been a technical success but it also implies that earlier lamp patents were failures compared to it. Once he had the innovation of high resistance, Edison abandoned the more complex low resistance lamps in the earlier patents. This pattern of minor improvements to a basic innovation continued and by the time Edison opened his first commercial power station in September 1882, he had applied for a further 82 patents relating to the incandescent lamp.

Compared to this intense inventive activity on electric lighting, his work on motion pictures appears almost incidental. Despite being a pioneer producer of motion pictures, Edison applied for only nine patents related to the technology. With controlling patents for the motion picture camera and the Kinetoscope for exhibiting them, he could also control the industry commercially, unlike electric lighting, where he was only one of many competing inventors with claims on the industry. Having that commercial control reduced the need to seek more patents.

The third group of patents intermingled with those for developmental stages and minor innovations are the defensive patents discussed earlier. To the extent that they achieved their purpose of discouraging competing inventions, they can be regarded as successes despite Edison not developing them into commercial successes.

The fourth group are patents for rather fanciful inventions. These include the Aerophone, a device that used a train whistle as a loudspeaker, which the New York Times feared would lead to "the complete disorganisation of society".[27,28] The Aerophone failed commercially because there was no demand for it although it appears to have been feasible technically. On the other hand, Edison also patented inventions that were unquestionably not feasible technically, like a flying machine that could never have flown (Fig. 9.2). In addition to these patented inventions, Edison also provided newspapers with a constant stream of wonderful devices that he was "going to invent" including one of his last, a device to communicate with the dead.[29] Since the patent office only enquires into whether an invention is novel, not whether it is technically feasible or even effective at achieving the inventor's claims, the awarding of a patent does not imply technical feasibility or likely commercial success.

The fifth group consists of the patents that, while not fanciful, were not commercially successful. Most notable among these are the patents relating to magnetic ore refining and alkaline batteries that Edison intended for electric automobiles. Magnetic ore refining caused Edison to loose around 300 million dollars in today's value, an almost catastrophic commercial failure, while the alkaline batteries failed because there was little interest in electric automobiles. Ever resourceful, Edison later successfully adapted his alkaline batteries to industrial applications and submarines.

---

[26]Edison. Electric Lamp.

[27]Improvement in Speaking Machine.

[28]New York Times. "The Aerophone." *New York Times*, 25 March 1878, 4.

[29]Forbes, "Edison Working on How to Communicate with the Next World."

The final group consists of patents issued to Edison but which other people developed. In the process of developing the incandescent lamp, Edison invented a high vacuum pump and this, in turn, led to a patent for preserving food in vacuum.[30] While this invention was of no commercial value to Edison at the time, the underlying technology is successfully used today. Edison also took out a patent in 1885 for wireless communication (discussed in Chap. 8).[31] Despite anticipating Hertz's wireless experiments, Edison did nothing more with the patent although Marconi was forced to buy the rights to it so he could exploit his own wireless telegraphy technology.

Edison's many patents are evidence of a remarkably inventive career but the small number of significant innovations indicates that the overwhelming majority of patents were either developmental stages, minor improvements to already successful inventions or unsuccessful commercially. This is not to suggest that only 12 out of 1084 patents were successes (1.1%). A more realistic figure of successful inventions would be something similar to the 10% estimated by United States Commissioner of Patents or over 100 successful inventions. It is still a remarkable result, even if it also implies many hundreds of failed patents. Thomas Edison was not a man to be deterred by failures, even if they were numerous.

9.6   Patterns in Edison's Patents

The patterns evident in Edison's patent output in Fig. 9.1 and Table 9.2 show an aspect of the man and his approach to inventing not evident from inspecting individual patents and inventions. One of the most conspicuous is the difference in the rate at which the younger Edison and the older man produced patentable inventions. Of his 18 significant inventions in Table 9.2, 15 were begun in the first 25 years of his career and only three in the remaining 38 years. This pattern is also evident in the rate at which he applied for patents, with first half taking only 18 years compared with 45 years for the second half.

The young Thomas Alva Edison (Al to his friends) was very much the working class boy who mixed with and employed men with a similar background to his own, and who married a young woman with a similar background. His early collaborators included Charles Batchelor, an English textile mechanic; John Kruesi, a Swiss-born clock maker; John Adams, a seaman turned inventor; and the Ott brothers, John and Fred, both machinists. (Both remained with him until Edison's death, Fred as Edison's personal machinist.) All these men came from the craft tradition and, like Edison, had little formal education. Consequently, they, like Edison, started with well-developed craft skills but little knowledge of science and mathematics. The young Edison's notebooks contain virtually no more mathematics than is relevant to

---

[30]Preserving Fruit.

[31]Means for Transmitting Signals Electrically.

a craft shop: quantities of parts for inventions and simple bookkeeping calculations. Soon after Edison started work on electric lighting, he began hiring better-educated associates including Francis Jehl; Francis Upton, a former student of Helmholtz[32]; Nicola Tesla, inventor of the alternating current electrical system[33]; Frank Sprague,[34] pioneer of electric traction; Reginald Fessenden,[35] who made the first radio broadcast of music and spoken word; and Henry Ford.[36]

The younger Edison worked with a small, close group of men who were essentially his equals but in 1887, when he built his much larger laboratory at West Orange, New Jersey, he had moved from being the first among equals to manager of a large invention factory employing up to 100 men. He had also remarried after the death of his first wife and, at the insistence of his second wife, was referred to as Thomas rather than Al. By 1903, Edison had also ceased to be the hands-on inventor photographed in Washington in 1878 and "seldom worked with his own hands. He had a mechanical man [Fred Ott] who did all the manipulating".[37]

The young Edison disliked the commercial pressures of manufacturing and marketing his inventions but the older Edison evolved into an industrialist,

---

[32]Francis Upton (1852–1921) Physicist, mathematician and Edison associate. Upton graduated from Bowdoin College and Princeton University, and then studied with Helmholtz at the University of Berlin (at the same time as Hertz). He joined Edison at Menlo Park November 1878 where he undertook theoretical work in connection with electric lighting. Edison nicknamed him "Culture"and so valued his contribution that he rewarded him with 5% of the royalties from the electric light patents.

[33]Nikola Tesla (1856–1943) Serbian American inventor, physicist and engineer. Tesla worked briefly for Edison in the 1882–3 but left after a dispute with Edison over money. Tesla developed the polyphase (alternating current) electrical system and induction motor, assigning the patents to George Westinghouse. Tesla's alternating current system eventually supplanted Edison's direct current in most electrical power applications.

[34]Frank J Sprague (1857–1934) American electrical engineer and inventor. Sprague joined the US Navy and was trained in engineering at the United States Naval Academy, Annapolis. He left the navy to join Edison at Menlo Park where he developed mathematical models for designing electrical utilities in new cities. Sprague left Edison to work on electric traction, establishing, with Edward Johnson, the Sprague Electric Railway & Motor Company. After the company merged with Edison's, Edison was credited with some of Sprague's inventions, a claim that Sprague vigorously contested. Sprague later became a member of the Naval Consulting Board over which Edison presided.

[35]Reginald Fessenden (1866–1932) Pioneer of radio communication. Fessenden was an Edison employee from 1886, later becoming professor of engineering at Purdue University and then the University of Pittsburgh. Fessenden made the first radio broadcast of music and spoken word in 1906.

[36]Henry Ford (1863–1947) American Industrialist and developer of mass production manufacturing. Ford was employed by the Edison Illuminating Company in Detroit from 1891 to 1899 becoming its chief engineer. Hughes claims that Ford's ideas for mass production developed from this period and that he used electricity as his metaphor for production. Like Edison, Ford distrusted experts. Ford became a close friend of Edison, whom he revered to the point of recreating Edison's Menlo Park complex by moving what was left of it to a site adjacent his Dearborn, Michigan plant. President Herbert Hoover dedicated the reconstruction on 21 October 1929.

[37]Rosanoff, "Edison in His Laboratory."

immersed in business entrepreneurship. A consequence of this was that his West Orange laboratory became the servant of his manufacturing business, concentrating on improvements to, and defence of, existing inventions rather than on developing new.[38]

The change from the younger Edison to the older man occurred around 1890. Figure 9.1 shows that Edison executed only ten patents from 1890 to 1895, a fraction of his previous output. In part, this can be attributed to Edison's disenchantment with the patent system, the struggle for control of his electric lighting ventures and involvement with his magnetic ore extraction venture. There were other significant events in this period, notably the acrimonious end in the early 1890s of his 20-year inventive partnership with Charles Batchelor and the death in 1899 of John Kruesi. With the loss of these two men, Edison could no longer go about inventing as he had in the 1870s and 1880s.

Whatever the cause or causes, the transition is so marked that several of Edison's biographers have proposed specific dates for the turning point. Conot nominates 6 October 1889, when Edison outwitted Vanderbilt interests to get control of his electric light company, transforming himself, in Conot's view, into an industrialist. "He had attained the pinnacle of the inventive world, and reached the zenith of his life".[39] For Hughes it was 2 January 1890 when Edison's first generating station in Pearl Street was almost destroyed by fire. "Edison's period of brilliance passed with the triumph at Pearl Street . . . After success at Pearl Street, other events in Edison's life added up to a watershed the gradual turn downward after the peaks of achievement."[40]

This is a pattern evident at the scale of Edison's whole life but we can find other significant patterns within the detail of Edison's patent output.

9.6.1 Electric Light and Power Patents

Edison's significance to the electric light is the consequence, not of his invention of the incandescent lamp but his invention of the first successful electric lighting system, the first electrical utility.[41] Describing his development of electric lighting as a system Edison said "The problem then that I undertook to solve was stated generally, the production of the multifarious apparatus, methods, and devices, each

---

[38] Andre J Millard, *Edison and the Business of Innovation*, Johns Hopkins Studies in the History of Technology (Baltimore: The Johns Hopkins University Press, 1990).

[39] Conot, *A Streak of Luck*, 286.

[40] Thomas P Hughes, "Thomas Alva Edison and the Rise of Electricity," in *Technology in America: A History of Individuals and Ideas*, ed. Carroll W Pursell, 2nd ed. (Cambridge, Massachusetts: MIT Press, 1990), 125.

[41] Friedel, Israel, and Finn, *Edison's Electric Light: Biography of an Invention*, 115.

adapted for use with every other, and all forming a comprehensive system."[42] Not only were many of his electric lighting patents part of this comprehensive system, but a significant number (57 out of 388 patents) are specifically system related with titles like "System of Electrical Distribution".

Historian Thomas Hughes extends the idea of electric lighting as a system by including a wide range of other aspects including social and political contexts to form "a seamless web".[43] Hughes further extends the notion of systems beyond electric lighting to claim that, "Edison's method was to invent systems rather than components for the systems of others" and that "During his long career as a professional inventor-entrepreneur, [Edison] turned to the invention of systems to such an extent that preference for systems can be identified as a salient characteristic of his approach."[44,45] It is important to note that when Hughes refers to a system, he means a large technological system, of which utilities like electrical power, telegraph and telephone are notable examples. This is however only one sense in which system can be used in relation to inventions and innovation. Chapter 4 develops a different view of a system and applies it to innovation but for the present discussion of Edison's patents, the term system refers to a large technological system.

Hughes's claim about Edison's system driven approach to inventing is important and influential claim and warrants closer examination.[46]

We can test Hughes's claim that Edison invented to build systems by looking for evidence of system building in his patents. If invented as a system there should be an identifiable pattern of Edison applying for patents on components of the system over a relatively short period with the patents collectively forming, in Edison's words, "a comprehensive system". Hughes characterises such systems as consisting of related parts or components connected by a network, or structure, often centrally controlled. He notes that the limits of the system are established by the extent of this control and that the purpose of control is to optimise the system's performance and to direct the system toward the achievement of goals. In the case of an electrical utility, the goal is to supply electrical energy to meet demand. Crucially, the interrelatedness of components of such a system means that each component influences other components in the system.[47]

When Edison started work on electric lighting, he had two existing models of utilities, to draw on: the existing and dominant gas lighting system and the emerging electric arc lighting system. Four years before he observed a wire carrying electrical current glowing incandescent in 1812, Sir Humphrey Davey had also produced an

---

[42]quoted in Hughes, *American Genesis*, 73.

[43]"The Seamless Web: Technology, Science, Etcetera, Etcetera," *Social Studies of Science* 16, no. 2 (1986).

[44]"Edison's Method."

[45]*Networks of Power: Electrification in Western Society 1880–1930*, 20.

[46]See, for example, Friedel, Israel, and Finn, *Edison's Electric Light: Biography of an Invention*. Israel, *Edison: A Life of Invention*. and the Edison s Book edition.

[47]Hughes, *Networks of Power: Electrification in Western Society 1880–1930*, 5.

intense light by striking an electric arc between two conductors.[48] Patents using these technologies for producing light from electricity were issued as early as the 1840s and by the time Edison started work on electric lighting in 1878, arc lighting was in a relatively advanced state of development, with Charles Brush installing arc lighting in the streets of New York in 1881.[49,50] Edison realised that while arc lighting was effective outdoors and in large spaces like theatres, it had serious disadvantages that limited its use in smaller rooms, notably the intensity of its light and the accompanying heat and fumes. When Edison began work on electric lighting he only intended to inventing an incandescent electric lamp but after identifying these problems with arc lighting, he determined to devise a way of "subdividing"electric light, that is, that each of his new incandescent lamps should give much less light than a single arc lamp.[51]

Figure 9.3 shows Edison's 388 electric light and power patents plotted to the same scale as Fig. 9.1. Although Edison applied for two patents in 1870 relating to electric motors, his main inventive effort on electric light and power did not begin until August 1878. Between then and the end of 1878 he executed six electric lighting patents, the annual number rising to a peak of 91 in 1882. The pattern of patent execution in Fig. 9.3 supports Hughes's claim that Edison was inventing to create a system that he had previously mapped out. It shows a prodigious burst of effort in the early 1880s that tapered off over the decade with occasional electric lighting patents over the rest of Edison's life. Because of the scattered pattern of these later patents, they, like his first two patents cannot be considered part of a cohesive effort at system building.

Despite Fig. 9.3 indicating Edison building an electric lighting system, he did not begin with a comprehensive concept of a system but came to it after several months. In April 1879, the *New York World* quoted Edison as saying "When I first started out [on electric lighting], I took into consideration only the lamp, but I soon became convinced that it was necessary to have a more powerful generator and feasible plan of subdividing the light".[52] In the interview, Edison explained how his original limited view expanded when he realised that for the lamp to work, he needed a way to distribute the power and for this, he needed to wire the lamps in parallel electrical circuits. Since parallel circuits resulted in a low resistance load on the generator, he then realised he needed to develop a new generator with low internal resistance to match the load. Edison's electric lighting system eventually expanded in steps like this to incorporate not only the generator and lamps, but all manner of devices

---

[48]Davy, *Elements of Chemical Philosophy Part I Vol I*, 85.

[49]Friedel, Israel, and Finn, *Edison's Electric Light: Biography of an Invention*, 7.

[50]Charles Brush (1849–1929) American inventor entrepreneur. On 18 December 1881, the Brush Company demonstrated New York's first public electric lighting system on Broadway between 14th and 34th Streets. Edison drew on Charles Brush's work on generators and arc lighting as he developed his own incandescent lighting system.

[51]TAED MBSB2:135.

[52]TAED MBSB2:135.

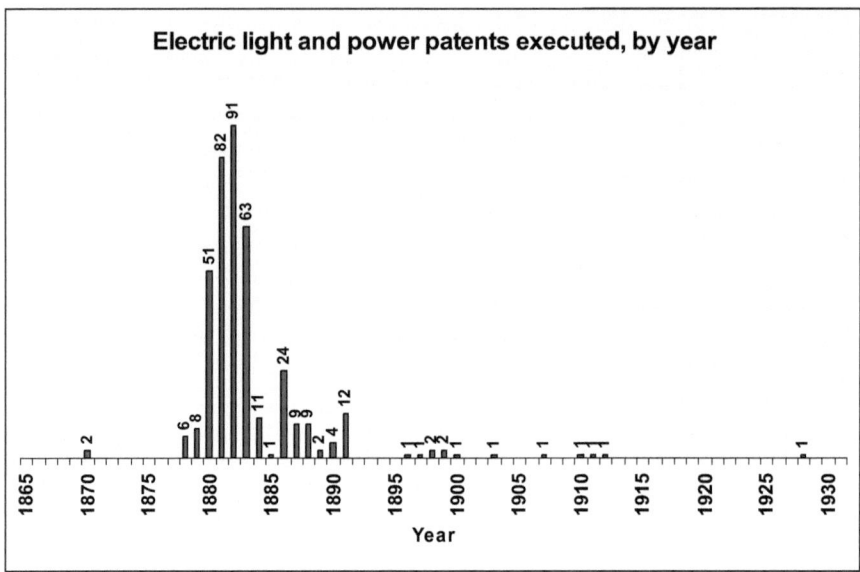

Fig. 9.3 Edison light and power patents, by year of execution

specific to electrical power distribution such as protective devices (circuit breakers) electric power meters, cables, current regulators, electric motors and a wide range of patents relating to manufacturing lighting apparatus, especially the manufacture of electric lamps.

A similar expansion process followed with components of the system. For example, Edison discovered that to achieve reasonable life with his chosen filament material (carbonised organic substances) he needed to extract residual volatiles from them and this, in turn, required a vacuum pump capable of achieving higher vacuums than currently available, leading him to patent improvements to vacuum pumps.[53]

Such development is consistent with Hughes's idea that systems are expanded to increase control with Edison expanding his system from a single component, the lamp, at first to include a generator and a plan for subdividing the light, and then expanded it further by adding more components.

In the midst of this effort on electric lighting, Edison applied for a patent for preserving fruit, an "economical method of putting up fruits, vegetables, and other organic substances in their natural condition, without cooking, for preservation in

[53]Thomas A Edison. Apparatus for Producing High Vacuums. US Patent 248,425, filed 29 March 1880, and issued 18 October 1881; Vacuum Apparatus [1]. US Patent 248,433, filed 31 January 1881, and issued 18 October 1881; Vacuum Apparatus [2]. US Patent 263,147, filed 30 August 1881, and issued 22 August 1882; Vacuum Apparatus [3]. US Patent 266,588, filed 6 December 1881, and issued 24 October 1882; Vacuum Pump. US Patent 251,536, filed 7 December 1881, and issued 27 December 1881.

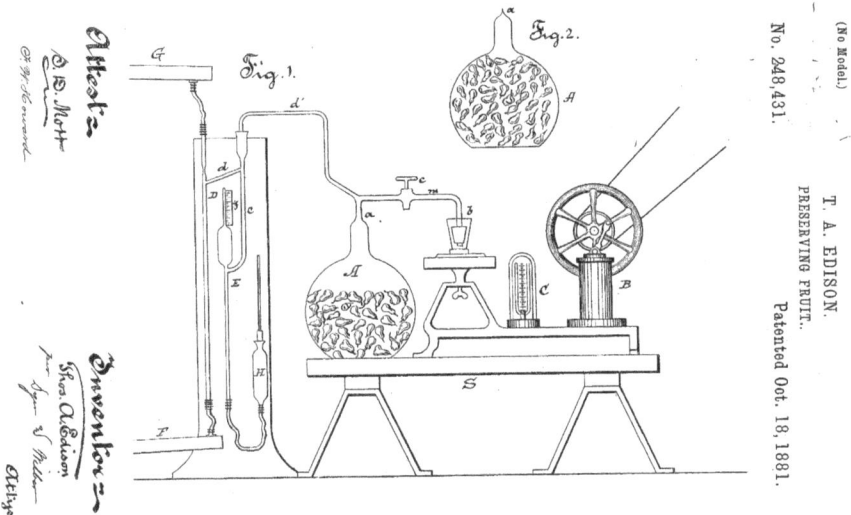

Fig. 9.4 Edison preserving fruit patent
Edison, Thomas A. 1880. Preserving Fruit. US patent 248,431 filed 14 December 1880, and issued
18 October, 1881

high vacuo".[54] To do this, the fruit to be preserved was placed in a glass vessel that is
then evacuated and sealed (Fig. 9.4).

At first sight, this might seem odd, unrelated to his then primary objective of
building an electric lighting system. Yet, despite it being radically different in
function from electric lighting, examination of its details and comparison with
other patents reveals that this is almost totally a product of the development of
Edison's electric lighting system.

On the same day that Edison's fruit preserving patent was issued, he was also
issued with two other patents involving vacuum producing equipment.[55] Fig. 9.5 is
the illustration that accompanied one of these patents and shows in the upper left
hand side three of Edison's lamps being evacuated. Comparing the vacuum
apparatus in the two patents shows they are almost identical. The principal
difference is not in the means for producing the vacuum (the pump) or what is
being evacuated (a glass bulb in both cases), but the contents of the bulb: fruit in
one and a lamp filament in the other. Edison had adapted the means for achieving
one function (producing incandescent lamps) to a completely different function
(preserving food). Edison's fruit preserving patent illustrates the way in which he
was constantly seeking to create new inventions either by extending the function of
other inventions or by creating new means for achieving existing functions.

---

[54]Preserving Fruit.
[55]Apparatus for Producing High Vacuums; Vacuum Apparatus [1].

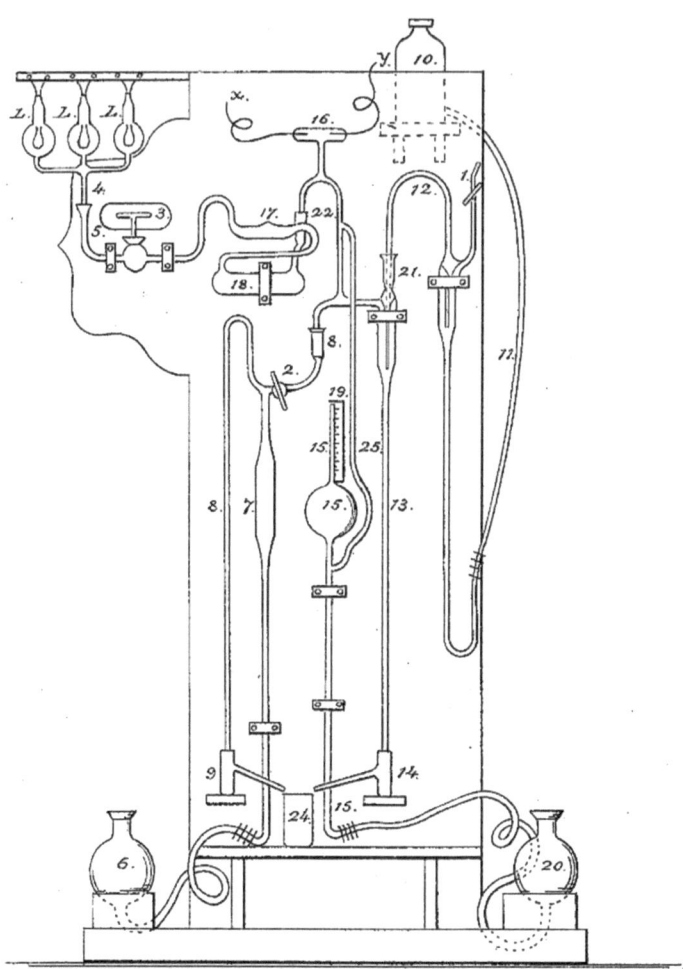

Fig. 9.5 Illustration from Edison's vacuum apparatus patent showing three lamps (upper left) being evacuated. (Apparatus for Producing High Vacuums)

9.6.2   Ore Processing, Mining, Cement and Concrete Patents

Edison's work on electric lighting branched into another area when he became interested in the possibility magnetically enriching low-grade iron ore. While working on electric lighting in 1880 he took out two patents relating to magnetic iron ore processing and three more the following year.[56] Other than briefly trying to apply his

---

[56]Magnetic Ore Separator. US Patent 228,329, filed 7 April 1880, and issued 1 June 1880; Magnetic Separator. US Patent 248,432, filed 6 August 1880, and issued 18 October 1881.

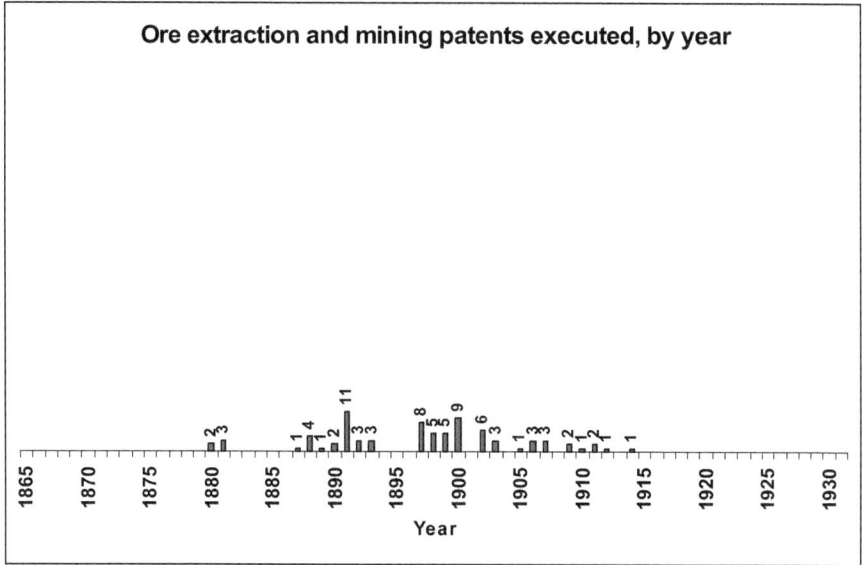

Fig. 9.6 Edison's magnetic iron ore extraction and mining patents

magnetic ore separator commercially to iron bearing sands in 1881–82, Edison did
not return to it until the end of the decade.

Figure 9.6 shows Edison's iron ore processing and mining patents. Edison
applied for a number of patents related to his magnetic extraction technology but
the problems he encountered in the heavy processes involved in ore milling also led
him to take out patents in related areas such a means for keeping abrasive dust out of
machine bearings and improvements to rollers for crushing ore.[57,58]

Edison made a major financial commitment to the ore processing project,
sinking a considerable amount of the wealth he earned from electric lighting into
it. When the venture collapsed in 1900, Edison adapted some of the inventions and
ore processing machinery to the industrial manufacture of Portland cement, the
basis of concrete. In this, as in iron ore processing, Edison developed a system of
related patents over a relatively short period (Fig. 9.7). To his existing iron ore
processing patents he added new ones related to cement and concrete, and adapted
their product, concrete, to a variety of uses to build a commercial enterprise. In so
doing, Edison become a pioneer of concrete, a construction material that was to
dominate the next century.

[57]Dust-Proof Journal Bearing. US Patent 472,752, filed 1 October 1891, and issued 12 April 1892.
[58]Roller for Crushing Ore or Other Material. US Patent 498,385, filed 1 October 1891, and issued
30 May 1893.

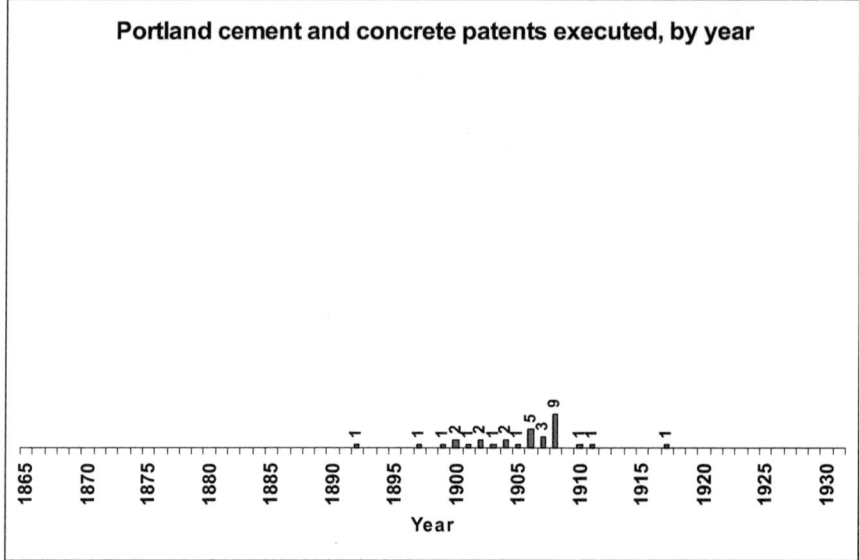

Fig. 9.7 Edison's patents relating to cement and concrete, by year of execution

His rotary kiln for producing Portland cement was so successful that it became standard in the industry and led to an increase in production of 230% between 1902 and 1907, being so successful that it led to an oversupply of the material.[59]

9.6.3   Phonograph Patents

The total number of electric lighting, magnetic ore extraction, and cement and concrete manufacture patents in Table 9.1 is around 480 or 44% of Edison's total patent output. The pattern evident in these patents supports the view that Edison invented in these fields to build systems. While they represent a significant proportion of his total inventive output, the question remains as to whether the other the rest of Edison's patents were the result of him inventing to build systems. We can begin to answer this by looking at Edison's favourite invention, the Phonograph.

Figure 9.8 shows Edison's 178 Phonograph patents by year of execution. Edison applied for his first Phonograph patent in December 1877, established the Edison Speaking Phonograph Company in April 1878 and soon after licensed the its manufacture in Britain and Europe.[60] In following this course, Thomas Edison, the entrepreneur was exploiting the fruits of the efforts of Thomas Edison, the inventor.

---

[59]Dyer and Martin, *Edison, His Life and Inventions.* 506.

[60]Edison. Phonograph or Speaking Machine.

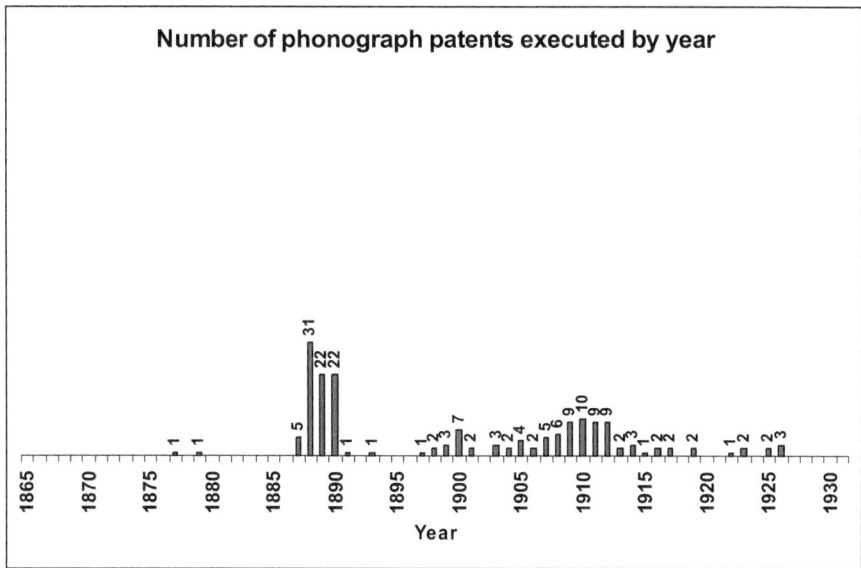

Fig. 9.8 Edison's Phonograph related patents by year of execution

This first Phonograph patent was followed by another in 1879, but it was not until November 1887, 10 years after the first, that Edison applied for his next Phonograph patents.[61] There followed a flurry of patent activity, with 36 patents executed in the next 12 months and 81 between November 1877 and February 1891.

One explanation for Edison's 10 years of inactivity on the Phonograph is that he had no need to do more since he had the controlling patent for sound recording and had established businesses to exploit it. Edison's monopoly ended in the 1885 when Alexander Graham Bell, his cousin, Chichester Bell, and Charles Tainter, patented a competing cylinder machine they named, with an unsubtle reference to Edison's invention, the Graphophone.[62] The Graphophone's inventors initially approached Edison with an offer to form a joint company but later established their own, the Volta Graphophone Company, and demonstrated a successful device in May 1887.[63]

The competition increased further when, in 1887, Emile Berliner applied for a patent for a disk machine with yet another name referencing the Phonograph, the Gramophone, which directly addressed weaknesses in the machine Edison's first Phonograph.[64] With the Bell-Tainter and Berliner patents, Edison's monopoly was at an end and his patent had only seven more years to run. The threat to Edison's

---

[61] Phonograph. US Patent 227,679, filed 29 March 1879, and issued 18 May 1880.

[62] Alexander Graham Bell, Chichester A Bell, and Charles Sumner Tainter. Reproducing Sounds from Phonograph Records. US Patent 341,212, filed 18 November 1885, and issued 4 May 1886.

[63] Israel, *Edison: A Life of Invention*, 280.

[64] Berliner. Gramophone.

monopoly of sound recording stirred him into action to defend his position. He did this expanding the Phonograph from a single device into a system of sound recording and reproduction.

Although we can see Edison building the Phonograph into a system in the pattern of his later patents, there is no evidence of system building patents in the first 10 years of the Phonograph's existence.

9.6.4 Motion Picture Patents

Even more striking than the pattern of Edison's Phonograph patents are those for motion pictures (Fig. 9.9). Edison was a pioneer of motion pictures, establishing the first purpose built studio, the Kinetographic Theatre, at his West Orange laboratory.[65] From a handful of patents, he created a group of companies to exploit his invention and in 1903, his company, Edison Film, produced what is claimed to be the first narrative movie, *The Great Train Robbery*. Edison continued producing motion pictures until 1918.

Edison built his motion picture business as a commercial system based on two patents executed at the end of 1891, one for a "Kinetographic camera" to film motion and the other for "Apparatus for exhibiting photographs of moving objects".[66] These became the controlling patents for motion pictures in the United States but, rather than adding inventions to build them into a system, Edison produced only seven other unrelated (i.e. non-system) patents over the next 22 years. With motion pictures, Edison was a system builder as an entrepreneur but not as an inventor, creating a commercial system to produce, market, distribute and exhibit motion pictures but not a system of inventions to go with it. That he did not build a system is somewhat surprising, because motion pictures offered Edison a range of opportunities in fields in which he had already demonstrated expertise. We could reasonably have expected, but do not, see Edison producing inventions based on his considerable knowledge of chemical processes, nor were there inventions for processing of motion picture film nor for production processes such as duplication of film prints and editing.

Edison may have built motion pictures into a commercial system but the pattern of his patents does not show Edison inventing to create a system.

---

[65]Despite becoming a pioneer of motion pictures, Edison almost missed the significance of motion pictures, initially believing they would be a parlour entertainment, like the Phonograph rather than mass audience entertainment like the theatre.

[66]Thomas A Edison. Apparatus for Exhibiting Photographs of Moving Objects. US Patent 493,426, filed 24 August 1891, and issued 14 March 1893; Kinetographic Camera. US Patent 589,168, filed 24 August 1891, and issued 31 August 1897.

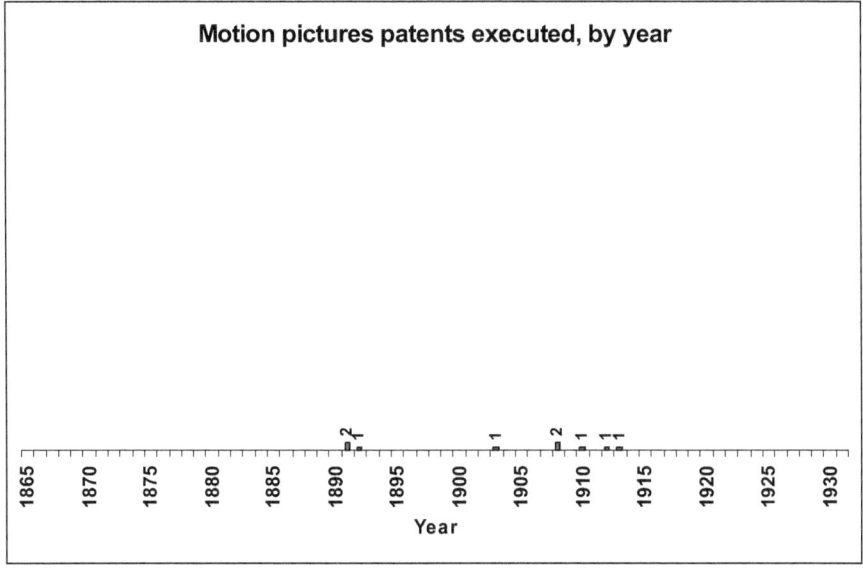

Fig. 9.9 Edison's motion picture patents by year of execution

9.6.5   *Battery Patents*

Edison's patents covering both primary and secondary batteries are plotted in Fig. 9.10. He applied for his first battery patent in 1872 and at least one battery patent every decade to the end of his life, eventually being issued with 107, 10% of his entire patent output. Edison's long interest in batteries is not surprising since they were a constant part of his environment and, even before he took up inventing, they powered his work as a telegrapher. Batteries also combined two of Edison's passions, electricity and chemistry.

In 1900, Edison became interested in the potential of electric vehicles. He realised that a critical weakness was the weight of conventional lead-acid batteries compared to their low storage capacity. To address this, he developed a version of the iron-alkaline battery. His battery patents of this period show Edison working to develop a system as he invented battery internal components like electrodes, devices for manufacturing batteries, filling them, protecting them and so on. While there is justification for claiming that Edison was inventing a system during this period, the same cannot be said for the rest of his battery patents that emerge almost at random out of other work rather than being developed to from part of a system.

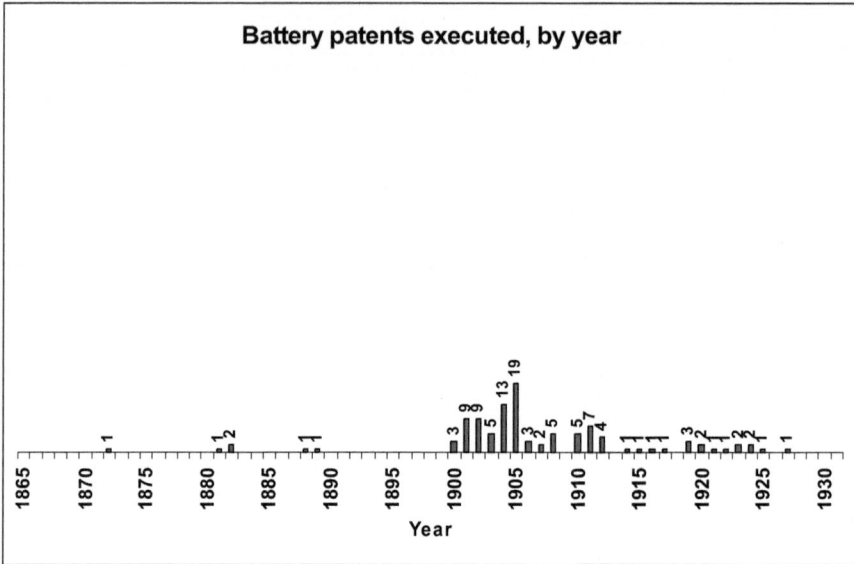

Fig. 9.10 Edison's battery patents by year of execution

9.6.6 *Electroplating Patents*

Like batteries, electroplating combined Edison's fascination with chemistry and electricity so it is not surprising that Edison explored electroplating several times during his career, applying for his first electroplating patent in 1882.

Figure 9.11 shows Edison's 30 electroplating patents spanning 51 years. While there is a small peak around 1905 coinciding with his alkaline battery development, there is no pattern that suggests a concerted effort at system building.

9.6.7 *Telegraph Patents*

A final, intriguing set of Edison's patents are those related to the telegraph. Of the 141 patents issued to Edison before the Phonograph, all but eight relate to telegraphy. Among these early patents are the duplex telegraph, quadruplex telegraph and his improvements to the printing telegraph, all crucial developments of telegraph technology. Two other significant inventions, the carbon microphone and Phonograph, were initially conceived as telegraph components. Edison's telegraph inventions were important financially and established his reputation as an ingenious inventor with telegraph companies. Through these inventions, he honed the techniques he later employed to develop the electric lighting and other systems.

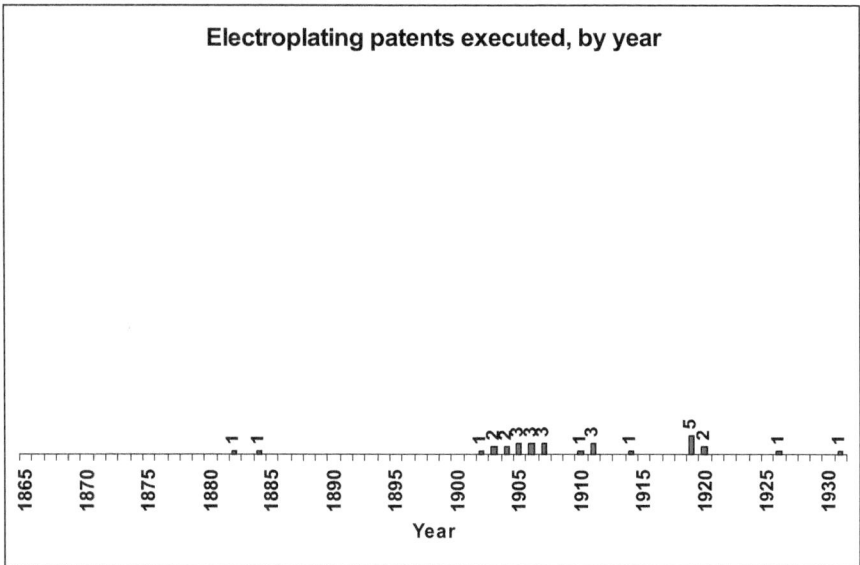

Fig. 9.11   Edison's electroplating patents

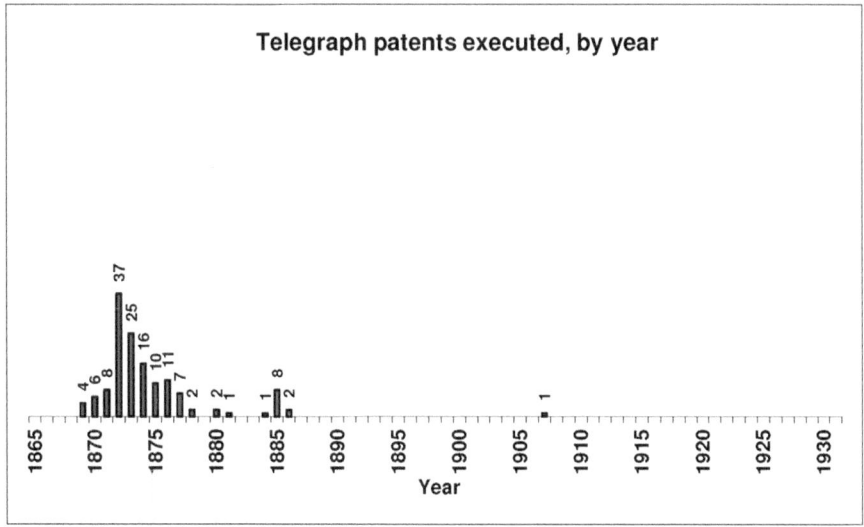

Fig. 9.12   Edison's telegraph patents

Figure 9.12 shows Edison's telegraph patents clustered around the early 1870s and might lead us to believe that he conceived them, like electric lighting as a system.

However, looking at his telegraph patents and the circumstances surrounding them it is apparent that he was inventing, in Hughes's words "components for the systems of others" and not his own system.[67] In fact, the extent to which he was inventing for others is apparent in his employment by Western Union expressly to produce inventions, notably the carbon microphone, to order. The cluster of tele-graph patents is simply the result of Edison at that time specialising in inventions for the telegraph.

9.7   Patents and Systems

It is apparent that Edison developed electric lighting, magnetic ore extraction, and cement and concrete as systems. It is also apparent that during specific periods Edison invented to build systems relating to batteries, the Phonograph, and electroplating. Viewed this way, around 750 of his patents can be seen as the result of Edison building these six systems. However, his remaining patents, about 30% of the total, were not created as part of systems and can be allocated to at least 50 other fields.

It is also apparent that, before he began work on electric lighting, Edison did not invent to build systems for himself but to create components for the systems of others, notably Western Union. At this early stage of his career, the few the exceptions to this principle are inventions that he intended to market himself such as the Inductorium and electric pen. Clearly, Edison made a crucial shift in focus from creating components for the systems of others, to building his own systems when he moved into electric lighting. This shift appears to be the consequence of his inventions being intended to supplant another system, gas lighting. When he began work on electric lighting, his initial motivation was to build a better incandescent lamp but he soon realised that there was far greater scope for creating something that could compete with gas lighting. He had been working for years with the telegraph, a large technological system and now hoped to compete with another large techno-logical system, gas lighting. Edison's electric lighting system with its central gen-erating plant, reticulation and use in individual premises closely resembled gas lighting utilities, which also produced gas in a central plant and reticulated it to consumers. As Friedel and Israel observe, "That an electric light would have to have practical power generation and supply networks behind it was not a novel concept."[68]

Having grasped the obvious model of gas lighting as a system and the potential of systems, Edison first built electric lighting as a system, then the Phonograph, magnetic ore extraction, motion pictures and cement and concrete as systems. What is surprising, given this, is that Edison failed to exploit the systems potential of industries in which he was a pioneer and which should have been commercially attractive to him as an inventor-entrepreneur. A prime example of this failure is that,

---

[67]Hughes, "Edison's Method."

[68]Friedel, Israel, and Finn, *Edison's Electric Light: Biography of an Invention*, 31.

after his initial commercial success with the telephone, he largely abandoned it. Similarly, although he obtained a small number of patents relating to motion pictures and developed them into a commercial system, he failed to pursue inventions in related technological fields such as film processing.

9.8 Inventing Systems

It is clear that the invention of systems was an important aspect of Edison's work after 1878 when he began to attack electric lighting. It is also clear from Edison's patents that he did many other things besides inventing to build systems. Unfortunately, the emphasis on using systems to view Edison's approach to inventing has resulted in questions about Edison's patents being approached as questions of how he selected what to invent, rather than how he invented. The systems approach has been of value in highlighting the ways in which inventors, like Edison, decided what to invent but has contributed little to understanding how they got from the selected project to the Patent Office.[69]

A further problem with this systems building perspective is the result of its emphasis on social aspects of large technological systems because it is sheds little light on inventor's personal drives. System building explains Edison inventing to exploit the monopolistic aspect of patent law but not why he chose to develop alkaline batteries in the early 1900s but not inventions relating to motion pictures.

9.9 Deciding What to Invent

It is apparent from the forgoing that the claim that Edison invented systems is actually a claim about how he selected invention projects, not a claim about how he went about the task of inventing. Thus, the question of how Edison invented has tended to be seen as a question of how he selected projects. From a system perspective, the salient aspect of Edison's approach was to select projects that could be built into such systems. While this may explain many of his patents, it is not appropriate for over 300 of his patents in at least 50 non-system areas, nor for

---

[69]There are some exceptions to this generalisation: Hounshell, "Elisha Gray and the Telephone: On the Disadvantages of Being an Expert." Reese V Jenkins, "Elements of Style: Continuities in Edison's Thinking," in *Bridge to the Future: A Centennial Celebration of the Brooklyn Bridge*, ed. Melvin Kranzberg, Brooke Hindle, and Margaret Latimer (New York, NY: New York Academy of Sciences, 1984)., Michael E Gorman and W Bernard Carlson, "Interpreting Invention as a Cognitive Process: The Case of Alexander Graham Bell, Thomas Edison and the Telephone.," *Science, Technology and Human Values* 15, no. 2 (1990). and Gorman, "Alexander Graham Bell's Path to the Telephone"; 1994. "Cognitive Psychology of Science." [Web Page]. http://www.iath. virginia.edu/~meg3c/psybull.html

patents in system related areas (like electric lighting) that occurred either early or late in his career when he was not working to build them into a system.

There is another explanation of how Edison chose invention projects, applicable to both system building and non-system building. The key is that Edison was an opportunist. He was an opportunist in many aspects of his life; when experimenting, in his commercial endeavours, with the press; and in promoting himself. When he came to select projects for inventions, Edison was also an opportunist, constantly on the lookout for new opportunities for invention. Viewed from this opportunistic perspective, building systems was just a source of opportunities but, even if a rich source, far from the only one. Another source of opportunities was failures illustrated through the invention of the carbon microphone and anomalous phenomena he attempted to exploit with Etheric force. Similarly, Edison was quick to seize the opportunity offered be even the meagre evidence as he did with his first Phonograph experiments.

Edison also looked for opportunities for adapting existing inventions to new functions and for finding new ways of achieving exiting functions. In the case of electric lighting, the development of better vacuum apparatus provided him with the opportunity to apply it to a completely different field, preserving food. With the Phonograph, Edison saw that a device originally conceived for recording telegraph messaged transmitted by telephone could be used to a completely different function, as a source of entertainment.

Put simply, the salient feature of Edison's project selection strategy was not the building of systems but opportunistic accretion: he added inventions when opportunities presented themselves. Some opportunities involved adding to his existing body of work to build systems but there are many cases of him by attaching new ideas to his existing body of work. In the case of electric lighting expansion of patents coincided with the development of electrical power as a system but expansion could also occur when he attached a patent for preserving fruit to electric lighting and the Phonograph to the telephone. Edison summed up his haphazard approach to systems when he wrote in 1879:

> My own practice for many years – a practice not adopted as the result of any plan or purpose, but arising from the natural habit of my mind, has been to study a subject for a time, and then taking out patents for such parts of a general system I may succeed in making. ... As completed they are a system based on different inventions or discoveries, some of which have been made years before the others, and as I went along, finding my way from day to day and year to year.[70]

The breadth of fields in which to sought projects to invent can be seen in the following list of potential projects that he drew up in May 1875:

Wanted

   1. A Method of making 'Malleable iron' out of cast iron
   2. Making cast iron as hard as steel & to have some of the same properties

---

[70]TAEB 1684.

3. A method of making 'sawdust' soluble to form a cheap substitute for Ebony Hard rubber or Celluloid
4. A cheap intense green equal to Aniline green without Iodine or Arsenic
5. An Electromagnet which does not require wire
6. The formation of organic substances for the decomposition of water under certain influence
7. A Kerosene or other oil lamp which burns without chimney & gives a bright light
8. A new Engraving process
9. A Galvanic Battery of equal constancy to Daniels but with an electromotive force equal or nearly equal to Grove
10. A substance which will not pass or pass but little P current & all of a N current
11. A New force for Telegraphic communication
12. Make soluble peroxidyzed paper with less acid than the Celluloid CO.
13. Cheap process for the extraction of low grade ores, decomposed earth like either Carb Ag. or H_2S Ag. ores.
14. Platina solution cheap & as delicate as Iodide of Potassium without drawbacks
15. A Detector for Gold & Silver at a distance
16. A Polarized Electromotograph
17. A Sexduplex Telegraph
18. A copying press that will take 100 copies & system
19. Cheap process of printing new.[71]

The tendency with "to do" lists is that some things get done and some do not. Edison's list is no exception. Edison pursued item 9 in his many battery patents, item 11 in his exploration of Etheric force and item 13 in his magnetic ore extraction venture. Many others, such as items 8, 16 and 17 represent further developments of Edison's existing inventions. Regardless of specific outcomes, the most notable feature of the list is the range of Edison's interests it reveals; even at this early stage in his career, (he was 28 at the time). While it is not surprising to find developments of telegraphy (item 17) and Edison's electric pen (item 8 and 18) in the list, it also reveals much wider interests including metallurgy and chemistry. Even at this stage, Edison saw himself as more than an inventor of telegraph equipment, despite it being his principal interest at the time. His telegraph work had suggested opportunities for branching out from the telegraph into new areas. He seized similar opportunities many times over his career, the telephone providing the opportunity to branch into sound recording with the Phonograph.

It is also notable that the list is headed "Wanted" and that all items except 11 ("New force for Telegraphic communication") imply inventions. For inventions, "wanted" means wanted by the marketplace, that is, these list items express functions to be achieved. Having identified these functions, Edison's objective as a professional inventor was to devise and patent means for achieving them. It is also notable that the inventions, with the exception of item 18, are for individual devices or processes. Only item 18 mentions the invention as part of a system. The idea of inventing to build systems is evident in the list but at this stage, it is the germ of an idea, not yet a driving force.

---

[71]TAED NS7501:1–3.

Chapter 10
The Edisonian Method: Trial and Error

10.1 "His Method Was Inefficient in the Extreme"

The day after Thomas Edison died, Nicola Tesla, who worked for Edison in 1882–83, was reported as saying, "His method was inefficient in the extreme, for an immense ground had to be covered to get anything at all unless blind chance intervened and, at first, I was almost a sorry witness of his doings, knowing that just a little theory and calculation would have saved him 90 per cent of the labour".[1] The method that Tesla derided was trial and error, a method that became so closely associated with Edison that it sometimes referred to as the Edisonian method.

Despite Tesla's view that Edison's use of trial and error was "inefficient in the extreme", it was central to Edison becoming America's most successful inventor and to the development of the first commercially successful incandescent lamp, Phonograph and carbon microphone. Not only did Edison use trial and error to produce a large number of inventions, he did it at a prodigious rate, peaking at 106 successful patent applications in 1882, the year that Tesla worked for him. Chapter 4 argued that Edison's success as an inventor cannot be explained simply as the result of dogged persistence, since the potential number of random solutions to the problems Edison solved is so great that "blind chance" as Tesla asserted, is inadequate to explain even one successful invention much less Edison's prodigious output.

Tesla's description of Edison's approach reflects a widely held view that trial and error is the crude and inefficient last resort of untrained and unskilled people. While it is certainly a common problem solving strategy for novices, it is also frequently mentioned by those with high levels of training and skill, including scientists working at the frontiers of research. For these scientists and innovative inventors like Edison, trial and error may be the only path to solving problems if no relevant theory exists.

---

[1]Quoted in New York Times, "Tesla Says Edison Was an Empiricist," 1.

© Springer Nature Switzerland AG 2019

I. Wills, *Thomas Edison: Success and Innovation through Failure*, Studies in History and Philosophy of Science 52, https://doi.org/10.1007/978-3-030-29940-8_10

Both novices and experts use trial and error because it is not a single technique but a group of related techniques that can be located along a continuum, the location of a particular instance depending on the way in which knowledge from past failures is used. At one end, the end associated with the novice, trial and error relies totally on chance; the novice blindly trying anything they imagine to be a potential solution and making no use of knowledge from previous failed trials. Moving along the continuum involves using more knowledge gained from failed trials. Effective users of trial and error are effective because of the way they exploit past failures and the knowledge gained from them. For example, they use failed trials to identify and discard future candidates that are unlikely to succeed without trying them, thus reducing the number of trials required. Effective users of trial and error also identify empirical relationships between failed results, repeatable patterns that relate changes in one or more parameters to an effect produced. These empirical relationships are referred to here as *systems of regularities* and are a valuable and ubiquitous form of knowledge with predictive value that can be used when no theory is available.

10.2   Edison's Use of Trial and Error

We saw in Chap. 2 that although Edison subscribed to the common view that failure is of value for identifying negative examples (what to avoid in future), in practice he used failure in more ways, and more sophisticated ways, than this. One of the most sophisticated in Edison's hands was trial and error.

From the outset of his work leading to the invention of the carbon microphone, Edison realised that to succeed in transmitting articulate speech he needed a means for varying electrical resistance (and hence current) in response to sound. In January 1877, he recalled having earlier discovered that vibration altered the electrical resistance of carbon.[2] Initially he believed that the resistance of carbon varied with pressure and so, after confirming this in principle using blocks of carbon, he and his associates began what was to be a long search for a more effective material and arrangement for mounting it.[3] In this search, Edison was hampered by an absence of relevant theory for the vibration sensitivity of carbon because, as Gorman and Carlson observe, "no one had yet developed a chemical theory that Edison could have used to identify a form of carbon with the electrical properties he wanted".[4] Edison's solution to this lack of relevant theory was to apply trial and error.

Working with the theory that the electrical resistance of carbon varied with pressure, Edison began searching for a material with superior pressure-resistance characteristics. To make the testing of candidate materials more consistent, Edison

---

[2]TAEB 3:844n1.

[3]TAEB 3:880.

[4]Gorman and Carlson, "Interpreting Invention as a Cognitive Process: The Case of Alexander Graham Bell, Thomas Edison and the Telephone.," 152.

devised an instrument that applied controlled pressure to a sample to measure its pressure-resistance behaviour.[5] In June 1877 after testing over 100 semiconducting substances and mixtures using this instrument, Edison settled on carbon granules mixed with silk fibres, naming the mixture "fluff".[6] He spent several months working with fluff and in August 1877 successfully demonstrated a fluff-based telephone between his Menlo Park, New Jersey, laboratory and Manhattan a distance of about 40 km.[7] Despite this success, he found the carbon granules separated from the silk over time, causing the microphone to lose its effectiveness. Edison then sought a way to fix the carbon to the silk and in this search tried varying the proportions of silk and carbon noting in August 1877, "even loose plumbago [mineral graphite] or equivalent will work this".[8] By December 1877 he had abandoned carbon-silk mixtures and was working with lampblack alone (finely divided carbon obtained as soot from smoky lamps). He subsequently found that by purifying the lampblack he could make the transmitted sounds much louder, purified lampblack becoming the basis of the first commercial carbon microphone.

This search for a suitable material led Edison from the observation of a novel phenomenon, the vibration sensitivity of carbon, to a commercial product via a series of trial and error searches, adopted because there was no suitable theory for predicting the electrical behaviour of carbon under pressure. Edison started with blocks of solid carbon then, using trial and error, sought a more effective material trying various forms of carbon alone and mixed with other substances. He also expanded his search by testing other semiconducting materials such as manganese oxide. Once he found an apparent solution (fluff), Edison proceeded to refine it again using trial and error. This led him to discover that the silk fibres were unnecessary pointing him in a different direction, the result of which was his final material, refined lamp black.

Although he used a succession of trial and error searches, each new search built on the results of previous searches, rather than simply adding more candidates to the original search. Edison also refined his searches by introducing a purpose-built instrument to test the pressure-resistance behaviour of candidate materials. While he may have been astute in applying trial and error, he was hampered by working for much of the time with what he later discovered to be an erroneous theory. At the end of 1877 Edison's associate Charles Batchelor wrote "Mr Edison also found out that plumbago does not alter its resistance by pressure as we at first thought but the increased pressure made better contact [between the carbon particles]".[9]

We can draw several observations from this process. Firstly, each trial and error search built on knowledge gained from earlier searches, so Edison's search was a number of essentially independent trials rather than one long series. Secondly,

---

[5]TAEB 3:981n1.

[6]TAEB 3:941.

[7]TAEB 3:1016n1.

[8]TAEB 3:1005.

[9]TAEB 3:1107.

Edison sought to increase his chance of success by broadening the range of semi-conducting materials tried rather than limiting himself to carbon and carbon mixtures. That is, he reduced the sensitivity of the search the initial assumption regarding carbon. On the other hand, his reliance on an erroneous theory (the pressure sensitivity of carbon) reduced the range of potential candidate materials and led him in the wrong direction for almost a year.

While Edison's laboratory records reveal a reasoned methodical approach, to outsiders his approach could appear otherwise. Richard Berger, a chemist who worked with Edison during World War I, characterised Edison's use of trial and error as "the try everything method" and recalled Edison writing on another chemist's report "Keep on trying. There are 17 million chemical compounds we haven't tried yet".[10] Even Charles Batchelor, Edison's closest associate for 20 years, expressed his frustration at times, writing in 1874 "After trying some 15262842981 different solutions of Brazilwood we've come to the conclusion that it is not worth a damn".[11] Despite such frustrations, Tesla and Berger misrepresent Edison's approach. As his invention of the carbon microphone illustrates, he did not simply "try everything" until "blind chance intervened" but systematically worked to reduce the number of candidates to be tried, increasing the likelihood of success.

10.3   Trial and Error Techniques

We can examine the varieties of trial and error through a simpler example than Edison's inventions. Suppose I need to open a door that I have not opened for some time and find I have misplaced the key. Fortunately, I know there is a duplicate key in a box with other odd keys, but I do not label the keys for security reasons, I have to resort to a trial and error to find the key. As discussed in Chap. 3, judging whether something succeeds or fails involves assessing it against success criteria. In our search for the correct key, the relevant success criterion is that it opens the lock: if a key opens the lock it is a success; if it does not, it is a failure.

10.4   Blind Trial and Error

The crudest trial and error approach, which I will call *blind trial and error*, is to pick a key at random from the box try it in the lock and if it does not open the lock put it back into the box, repeating this until we find a key that opens the lock. This approach is blind because it learns nothing from the results of failed trials. Clearly,

---

[10]Richard G Berger, "With Edison's Insomnia Squad," *Modern Mechanix*, April 1934, 52.

[11]Quoted in Conot, *A Streak of Luck*, 71.

this is "inefficient in the extreme" because if we put failed keys back into the box with the untried keys, we are likely to try some more than once. We have made no use of knowledge from failed trials, the number of potential candidates remaining constant throughout.

10.5   Simple Trial and Error

A more refined approach is to set aside those keys that do not fit, each unsuccessful trial reduces the number of future trials. I refer to this approach as *simple trial and error*. It is what is commonly understood as trial and error, and seems to be what Tesla and Berger describe Edison using. It also fits Edison's conclusion from his "crowning experiment" anecdote.

While this simple try-and-set-aside approach may work with a single item like a key, with complex inventions the interactions between components make it an improbable formula for success. Edison's notebook entries on the carbon micro-phone show that he tried more than 150 resistance materials, dozens of diaphragm materials and a large number of arrangements for holding them. At 150 materials, this was a modest trial and error search by Edison's standards as he tested over 6000 substances in his search for an incandescent lamp filament and 9000 while developing his alkaline battery.[12] If Edison had tried 150 variable resistor materials, 20 diaphragm materials and 20 arrangements for connecting them, he would have faced 84,000 possible combinations to test using simple trial and error. Add to these variations in factors like the thickness of the diaphragm and purity of the resistance material and the chance of finding a successful combination using random trials is improbably low. Edison's prolific output of successful inventions indicates that he was doing something more than simple trial and error.

10.6   Informed Trial and Error

Simple trial and error is inefficient because it uses so little of the information from failed trials, building no more than a list of failures, in Edison's words things that "can't be done that way".

In some cases, nothing more than simple trial and error is possible because trials yield no information other than success or failure, for example "brute force" password attack, which yields no information other than whether a password tried works or does not. However, on looking at my box of keys after a few tries I realise that I can do better than try-and-set-aside. I can speed up the search by analysing the keys that failed and so derive some general idea of keys that are likely or not to

---

[12]*Edison, His Life and Inventions*. 262. ibid., 2: 615.

succeed. These are success clues. I notice that keys above a certain size are too big to fit into the keyhole and others have a shape that prevents them from fitting. I also notice that other characteristics like material are irrelevant in predicting whether a key fits the keyhole. Using this knowledge, I can examine the untried keys, selecting likely candidates and setting aside those that do not. By adopting this course, I might be able to discard most of the untested keys without needing to try them.

Using the acquired knowledge serves to reduce the number of trials required, a process I refer to as *informed trial and error*. An amount of simple trial and error often remains after the application of informed trial and error but it is applied to a reduced the number of candidates. Thomas Midgley, who discovered tetraethyl lead, a fuel efficiency booster for gasoline engines, said of his own use of trial and error, "The trick is to turn a wild goose chase into a fox-hunt".[13,14] Blind trial and error and informed trial and error are Midgley's wild goose chase and fox-hunt.

In the light of these examples, trial and error becomes not as a single method, but as a continuum running from blind trial and error, through simple trial and error, to various degrees of informed trial and error, the position on the continuum depending on how the way in which information from failed trials reduces the number of future trials. Simple trial and error, the form of trial and error commonly seen as the sole trial and error method, makes use of this information in only one way: to produce a list of failed candidates to be avoided in future. In the extreme, analysis of failed trials in informed trial and error might eliminate all but one candidate, ending the trial and error process. In the box of keys illustration, I might find that by examining the shape and size of each key and comparing it to the hole in the lock I can reject all but one key.

10.7   Using Trial and Error Instead of Theory

Analysis of failed trials can produce systems of regularities. In the key search example, a regularity is the discovery that that keys of a certain shape fit into the keyhole, and so are potential candidates, while keys with other shapes do not fit and can be rejected. Another regularity is Edison's discovery that the more he purified carbon black the more effective his carbon microphone became. These systems of regularities are quite specific to their respective origin in locks and microphones and have no claim to universal application. In contrast, Edison's erroneous belief that the resistance of carbon varied with pressure related to the fundamental properties of a material and potentially had more universal application beyond microphones.

---

[13]quoted in Hughes, *American Genesis*, 52.

[14]Thomas Midgley (1899–1944) American mechanical engineer, chemist and inventor. Midgley is best known for discovering tetraethyl lead, a fuel efficiency booster for gasoline engines, and dichlorodifluoromethane, originally marketed under the trade name Freon, a chlorofluorocarbon (CFC). Dichlorodifluoromethane was the first refrigerant that was both safe and effective.

Steinle argues that what we refer to as scientific laws are the most general regularities, part of a coherent system of regularities such that "any particular effect can be attributed a definite place within the system connected to the core effects by a chain of intermediate effects and thus be explained by or 'reduced' to the general laws. Thus the system of regularities gains explanatory power".[15] At their most basic, systems of regularities need have no explanatory power. They express a repeatable pattern that connects parameters with their effect without necessarily implying anything about why any parameter affects the effect. Despite this, systems of regularities do share an important characteristic of theories because both have predictive capacity. Knowing how a group of parameters affects an effect allows us to predict the effect within limits.

Examples of such systems of regularities from a different field are the empirically derived "maxims" relating to water wheels proposed by the eighteenth century engineer John Smeaton. Two of Smeaton's maxims are:

> In a given undershot wheel, if the quantity of water expended be given the useful effect is as the square of the velocity,
>
>   In a given undershot wheel, if the aperture whence the water flows be given the effect is as the cube of the velocity.[16]

In discussing these maxims, Layton comments "Neither could be classed as laws of nature; they were law-like statements about man-made devices. They were not logical deductions from the science of mechanics; they constituted the germ of a new technological science".[17] Smeaton's maxims are systems of regularities. Using them, nineteenth century engineers were able to predict waterwheel behaviour. Later they were shown to be theoretically derivable from theories of mechanics but as systems of regularities, they are not dependent on this theoretical justification for their predictive value.

Another example of a system of regularities that had a profound effect was James Watt's invention of the steam indicator in 1793. By this time Watt had made significant improvements to steam engines but was facing competition from other manufacturers. In order to keep ahead of them, he needed to improve the efficiency of his engines. He devised a simple indicator consisting of a piston pressing against a spring, the amount of movement indicating the pressure. In 1796 Watt's draftsman, John Southern, added a sting and pulley attached to the steam engine piston and moved a drawing sheet back and forth. Figure 10.1 shows a later commercially produced indicator using Watt's principles.

In Southern's improved indicator version, the drawing paper moved up and down in response to pressure and sideways in response to piston movement and volumetric

---

[15]Steinle, "Experiments in History and Philosophy of Science," 421.

[16]quoted in Edwin T Jr. Layton, "Mirror-Image Twins: The Communities of Science and Technology in 19th-Century America," *Technology and Culture* 12 (1971): 566.

[17]Ibid.

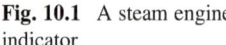

Fig. 10.1 A steam engine indicator

displacement while a pencil traced this pressure-volume relationship to produce the kind of figure visible in Fig. 10.1. Using trial and error, Watt and Southern discovered that, for a given engine, as the area of the figure increased, the amount of work the engine produced also increased for a given amount of fuel burned. That is, the larger the area the more efficient the engine. Armed with this, Watt could vary factors associated with the engine to determine how they affected work produced and so further improve engine efficiency. The system of regularities inherent in the indicator was so valuable to Watt that indicator remained a trade secret until 1822, 3 years after his death.

The indicator may have been the product of trial and error and employed in other trial and error experiments but it is an expression of fundamental physics, being an empirical form of the First Law of thermodynamics, first expressed by Rudolf Clausius in 1850. The area in the figure the indicator draws represents the mathematical product of pressure and volume or mechanical work.[18] In this example,

---

[18]Derived thus:

Area inside figure = pressure ∗ volume

Watt's system of regularities was an expression of a fundamental principle of physics not discovered until decades after his death.

It is the predictive capacity of systems of regularities that makes them so valuable in the efficient use of trial and error but their application is bounded by their empirical origin. While I can use a system of regularities to predict the likely success of a new key in a particular lock, its limited empirical basis means I am not justified in extending the generalisations beyond that specific lock. I cannot say on this basis that all keys of a particularly shape fit all locks of a particular kind.

We can be justified in using the predicative capacity of systems of regularities when interpolating within the range of the original empirical data but the same degree of confidence cannot be applied to extrapolation.

Despite this caution, there are some generalisations we can draw and apply to new situations. When we find that shape and size of keys are predictors of success and failure we learn that we can use these parameters as success clues in other lock situations, even if the specific shape and size we have discovered for this lock are not applicable. That is, the generalised knowledge we have acquired relates to classes of success clues. In the search for a successful solution to my key problem, I have only one success criterion: does the key open the lock, but along the way, trial and error has yielded several success clues that can help to identify the candidates that are more likely to succeed. Not only must the key open the lock, the relevant success criterion, but a successful key must also satisfy relevant success clues such as being of a specific shape and size. Often superficial features like size and shape are predictors of success but not themselves success criteria.

This process of identifying success criteria and success clues is evident in Edison's invention of the carbon microphone. When he began work, Edison's principal success criterion was that his invention should transmit articulate speech. As development proceeded, he refined this objective by adding success clues, for example that for speech to be articulate, the microphone must also faithfully transmit vowels and sibilants.[19] Using these success clues, he could predict whether a candidate microphone configuration would succeed or fail at transmitting articulate speech by testing it on vowels and sibilants in isolation. If it failed to transmit vowels, it would also fail to transmit articulate speech.

Since the systems of regularities produced by trial and error, like theories, have predictive capacity, they can be used as substitutes for theories when no relevant theory is known to the user. This is one reason why trial and error is associated with novices, since a characteristic of novices is that they do not know relevant theory. Part of Edison's approach to inventing in new fields was to build such systems of regularities into a systematic body of knowledge to fill the place of theories.

Pressure = force / area

Area inside figure = force / area * volume = force * length

Mechanical work = force * displacement = force * length

So area inside figure = Mechanical work

[19]TAED TI1:38, TAEB 3:882.

Reflecting on Edison's use of such knowledge, one of his associates, the mathematician Francis Upton, observed, "One of the main impressions left upon me after knowing Mr. Edison for many years, is the marvellous accuracy of his guesses. He will see the general nature of a result long before it can be reached by mathematical calculation".[20]

10.8   The Use of Trial and Error in Exploratory Experiments

Much of Edison's use of trial and error fits Steinle's concept of exploratory experiments discussed in Chap. 8, the epistemic purpose of which is to establish not theories but systems of regularities.[21] Steinle emphasises that "exploratory experimentation [is characterized by] the systematic variation of experimental parameters. The first aim here is to find out which of the various parameters affect the effect in question and which of them are essential".[22] Exploratory experiments involve varying many experimental parameters to determine which are indispensable. This knowledge is then used to establish stable empirical rules and means for representing them. Steinle contrasts exploratory experiments with theory driven experiments, which start with a theory and seek to reduce rather than expand the experimental arrangement so that there are "nearly no possibilities of varying the experimental conditions and the arrangement allows exactly the anticipated [effect] but no other".[23]

Experimenters beginning the exploratory experimentation phase on a particular effect cannot be certain which parameters affect the effect nor do they have relevant theory since developing that is an objective of the process. Consequently, in exploratory experiments the process of identifying which parameters are relevant and which are not is inherently one of trial and error. Trial and error is fundamental to this aspect of exploratory experiments (sorting relevant and non-relevant parameters) even if the variation of experimental parameters is systematic and not subject to trial and error. Experienced experimenters are more effective in exploratory experimental searches because they are able to exploit knowledge acquired from previous searches and so identify promising candidate parameters to include in the trial and error parameter searches.

We can add this use of trial and error (the sorting of relevant from irrelevant parameters) to other more obvious uses of trial and error in experimental science. We saw in relation to Etheric force (Chap. 8) that Edison's weakness as a scientist was to

---

[20]quoted in Dyer and Martin, *Edison, His Life and Inventions*. 620.

[21]Friedrich Steinle, "Entering New Fields: Exploratory Uses of Experimentation," *Philosophy of Science* 64, no. Proceedings (1997); "Experiments in History and Philosophy of Science."

[22]"Experiments in History and Philosophy of Science," 419.

[23]"Entering New Fields: Exploratory Uses of Experimentation," S.67.

concentrate almost exclusively on exploratory experiments, stopping when he had enough systems of regularities for his immediate needs rather than proceeding to the theory driven experiment stage from which general laws could be developed.

Despite denigrating Edison's approach, in his own work Tesla also used to a trial and error, notably when he and Charles Scott were trying to develop an induction motor for streetcars.[24] Later, when he had a successful induction motor, he indirectly acknowledged his use of empirically derived systems of regularities as opposed to theory, saying that "although my motor is the fruit of long labour and careful investigation, I do not wish to claim any other merit beyond that of having invented [the motor], and I leave it to men more competent than myself to determine the true laws of the principle and the best mode of its application".[25]

10.9   Transmitting Trial and Error Knowledge

Systems of regularities are transmissible from one person to another. Some are amenable to succinct expression like Smeaton's maxims while others can be transmitted in more complex forms such as tabulated empirical results, graphs, and sets of "dos and don'ts". In the key search example, systems of regularities can be expressed as statements like "this kind of lock requires this shape of key". This kind of knowledge is referred to sometimes as tricks of the trade or trade secrets because they can be transmitted from an expert to a novice, for example from a master locksmith to an apprentice.

This kind of knowledge also forms a significant part of the content of engineering handbooks where it is presented alongside theoretically derived knowledge. Handbooks dealing with fluid flow, for example, present theories such as the Bernoulli equation alongside graphical presentation of empirical data like the Moody Diagram.[26] The Moody Diagram is a graphical representation of the complex relationship between friction, Reynolds number (a function of velocity, conduit dimensions and fluid viscosity) and conduit roughness, and itself combines both theoretical considerations and empirical data.[27]

According to Polanyi, "we remain ever unable to say all that we know".[28] That is, there is an important part of what we know personally which we cannot articulate

[24]Ronald R Kline, "Science and Engineering Theory in the Invention and Development of the Induction Motor, 1880–1900," *Technology and Culture* 28, no. 2 (1987).

[25]quoted in "Tesla and the Induction Motor," *Technology and Culture* 30, no. 4 (1989).

[26]See, for example, ASHRAE, *ASHRAE Handbook - Fundamentals (SI Edition)* (Atlanta, GA: American Society of Heating, Refrigerating, and Air-Conditioning Engineers, Inc., 2009), 3.1–3.14.

[27]Lewis F Moody, "Friction Factors for Pipe Flow," *ASME Transactions* 66, no. November (1944): 671, 72.

[28]Michael Polanyi, *Personal Knowledge: Towards a Post-Critical Philosophy*, 1962 corrected ed. (London: Routledge, 1958), 95.

either to ourselves or to others. Polanyi calls kind of knowledge tacit knowledge. Since systems of regularities are transmissible from person to person, sometimes in very explicit form, they are not normally tacit knowledge.

Collins extends Polanyi's concept of tacit to include "knowledge or abilities that can be passed between scientists by personal contact but cannot be, or have not been, set out or passed on in formulae, diagrams, or verbal descriptions and instructions for action".[29] In examining the application of tacit knowledge to science, Collins discusses the attempts by British laboratories to replicate the TEA laser after it had been successfully demonstrated in Canada.[30] He notes that although the British possessed explicit (written) accounts of the successful Canadian device, they lacked other knowledge possessed by the Canadian researchers. In time, they acquired this missing knowledge, some by personal contact with the Canadians and some through their own trial and error experimentation. While part of what they learned fits Polanyi's concept of tacit knowledge, Collins gives examples of an experimenter articulating what had been learnt such as "[there is] no limit to how short [capacitor leads] should be, just as short as possible" and "beware my lad when we go to short, high voltage stuff – transients and odd things creep in".[31] Although expressed in somewhat vague terms, this knowledge is articulable. Both statements are expressions of systems of regularities, the first being an explicit expression of a success clue. In the light of this, it appears that Collins's concept of tacit knowledge can be more accurately understood as including both tacit knowledge in Polanyi's sense and transmissible systems of regularities.

Like Collins's TEA Laser scientists, Edison's experiments yielded both tacit and explicit knowledge, the latter often in the form of systems of regularities. Like Collins's experimenters, the systems of regularities Edison discovered were often expressible only in vague terms. In March 1878, Edison encountered a problem with a small piece of rubber used as a spring in his carbon microphone. To overcome the problem he tried progressively stiffer metal springs eventually replacing the rubber component with a rigid brass tube. Edison described his discovery in a letter to a colleague "I kept on substituting spiral springs of thicker wire, and as I did so I found that the articulation became both clearer and louder".[32] In doing so, Edison expresses a regularity similar to the TEA laser regularity about capacitor leads: the stiffer the spring the better the articulation of transmitted speech.

We also find many references to systems of regularities in Edison's patents, often as statements beginning "I have discovered" or "I find that". In an early electric generator patent, Edison says, "I have discovered that by combining with the field-

---

[29]Harry Collins, "Tacit Knowledge, Trust and the Q of Sapphire," *Social Studies of Science* 31, no. 1 (2001): 72.

[30]*Changing Order: Replication and Induction in Scientific Practice* (Chicago: University of Chicago Press, 1992), 51–78.

[31]quoted in ibid., 60–61.

[32]TAEB 4:1251n3.

of-force magnet and the circuit through the same a shunt open and closed periodically the dynamic effect is greatly increased".[33] This states a system of regularities. Edison proposes no explanation of why it occurs, no indication of how much the effect increased or how that increase compares to other methods of producing an increase. Statements like this may be articulable, not tacit, but the reader of Edison's patent gains little that can be applied to new situations, partly because of the vagueness of the statement and partly because Edison must have possessed, but did not disclose for commercial or other reasons, the rest of the empirically derived knowledge that is implied by the statement. In most cases, systems of regularities are mentioned only coincidentally in Edison's patents but in some, they are central to his claim of novelty. Of the 200 words of a very brief patent relating to chemical telegraphs, Edison devotes 140 words to describing systems of regularities.[34]

While Edison may have been able to use the trial and error derived systems of regularities, the difficulty of transmission could create problems for those who worked with him. Edison may have been able to design an electric generator using systems of regularities he had learned but only he could do it and he did not, or could not, transmit this knowledge to others.

Not only were Edison's systems of regularities difficult to transmit but there is evidence that even when they were potentially transmissible, Edison intentionally chose not to. In the case of his generator patent, Edison had to disclose enough knowledge to gain the patent but not enough to help competitors. Edison had a commercial reason for doing this but he also withheld information from new people who joined his laboratory team, his intention being that they learn it themselves. This approach worked for some, like Berger, but Rosanoff, another chemist, complained that Edison intentionally withheld laboratory notebooks containing his predecessor's experimental results, Edison's objective being to force him to repeat the experimental learning process that his predecessor had been through.[35] Eventually Rosanoff solved the problem Edison had set him (finding a more effective wax for Phonograph recordings) but commented that it came only "after a year of Edisonian blind groping that had led nowhere except to my having learned to think waxes".[36] That Rosanoff came to a solution after having learned to "think waxes" suggests that he did so by acquiring through trial and error, the systems of regularities and tacit knowledge needed to point to a theoretical solution to the problem. In doing so, he followed what Edison did with the first Phonograph demonstrated to *Scientific American* in December 1877. It appears that it worked, to Edison's surprise, because the trial and

---

[33]Thomas A Edison. Improvement in Dynamo-Electric Machines. US Patent 219,393, filed 10 July 1879, and issued 9 September 1879.

[34]Improvement in Solutions for Chemical Telegraphs. US Patent 168,466, filed 26 January 1875, and issued 5 October 1875.

[35]Rosanoff, "Edison in His Laboratory," 404.

[36]Ibid., 416.

error experiments conducted by Edison and Batchelor over the preceding few months enabled Edison to "think Phonographs".

10.10   Scientists' Use of Trial and Error

In accepting the Perkin Medal for distinguished work in chemistry, Thomas Midgley gave an account of his search for a fuel additive to control knocking in internal combustion engines:

> [In the search for an antiknock agent] the following determinations were arrived at by the Edisonian method:
>
> 1. Elemental iodine dissolved in motor fuel in very small quantities greatly enhances the antiknock characteristics of the fuel (the basic discovery).
> 2. Oil soluble iodine compounds had a similar though modified effect.
> 3. Aniline its homologs and some other nitrogenous compounds were effective though their effectiveness varied over a wide range depending upon the hydrocarbon radicals attached to the nitrogen.
> 4. Bromine carbon tetrachloride nitric acid hydrochloric acid nitrites and nitro compounds in general increased knocking when added to the fuel and air mixture.
> 5. Selenium oxychloride was extremely effective as an antiknock material.
> 6. A large number of compounds of other elements had shown no effect.
>
> With these facts before us, we profitably abandoned the Edisonian method in favour of a correlational procedure based on the periodic table. What had seemed at times a hopeless quest covering many years and costing a considerable amount of money rapidly turned into a "fox hunt". Predictions began fulfilling themselves instead of fizzling.[37]

Midgley's account illustrates both the value of trial and error and its pitfalls. Echoing Rosanoff's experience solving Edison's wax problem, it appears that Midgley's determinations yielded many failures, a few successes but, most significantly, systems of regularities pointing the direction of an analytical rather than trial and error approach. The value of trial and error in this instance lay not in identifying the large number of agents that had no effect or the few agents that had some effect, but in helping Midgley identify a more profitable analytical approach, one he had not considered before starting with trial and error. Midgley also recognised the need to abandon trial and error at some point, although his tone suggests that he thought it might have been better to abandon it earlier than he did.

In describing Edison's method as "inefficient in the extreme" Tesla implied that Edison used trial and error because his lack of education meant he was unable to use "a little theory and calculation". While it is true that Edison attended school for only 3 months, he was not uneducated as he received most of his early education at home from his mother who had been a teacher. Compensating for a lack of further education (he started work at 12), Edison was a voracious reader from his teens.

---

[37]Thomas Jr. Midgley, "From Periodic Table to Production," *Industrial and Engineering Chemistry* 29, no. 2 (1937): 242.

Later he kept abreast of current scientific literature and employed better-educated specialists to coach him in areas where his knowledge was limited. He may not have had much formally education but he was far from a simple, uneducated country boy.

While Tesla's reasoning has some basis in Edison's case, there are many references to the use of trial and error in present-day scientific literature by people who are very able to apply theory and calculation. A search for "trial and error" in an online scientific database yields thousands of citations in a broad variety of fields. Just one example is a comment by a researcher that "As so often happens in human endeavour researchers were forced to use the Edisonian Method, inspired trial and error, and eventually as luck had it they found some success".[38] Trial and error may have been necessary but there is no need to explain its success through notions of inspiration or luck.

In discussing the problem faced by scientists confronted by novel observations, Gooding observes, "with respect to novelty, *everyone* is a novice".[39] Developing this observation, Nickles argues that at the frontier of research, experimentalists are in some respects and of necessity, novices rather than experts. In doing so, Nickles uses the terms expert and novice relatively. It is not that at the frontier of research trained scientists are no better than those who are new to the field, rather they are novices in the sense that "they have not yet brought new phenomena' ideas, techniques, etc, under complete intellectual and practical control" and that "Sometimes they are experts at being frontier novices!"[40] Notwithstanding being new to the circumstances to which their research leads them Nickles adds, "Scientist novices-at-the-frontier are normally much better off than complete lay novices would be, since their training provides a large stock of resources and previous experience to apply to the new problems".

In a similar vein, Galison observes, "Expertise developed in prior work pays off in the specialised labour needed to analyse the behaviour of specific equipment or disturbing effects understandable from other branches of physics".[41] For Steinle "previous experience in the field or in related ones provides some ideas about where to start i.e. about what might be promising candidates for being relevant parameters and what not".[42] Applying Nickles's observation to Collins's TEA laser study, we could describe the Canadians in as experts and the British researchers as relative novices. Both groups were anything but "complete novices" as both possessed very high degrees of relevant knowledge and experience in the field. However, relative to the Canadians the British were "scientist novices-at-the-frontier".

---

[38]Eli Yablonovitch, "Photonic Crystals: Towards Rational Material Design," *Nature Materials* 2 (2003): 648.

[39]Gooding, "How Do Scientists Reach Agreement About Novel Observations?," 208.

[40]Thomas Nickles, "Justification and Experiment," in *The Uses of Experiment : Studies in the Natural Sciences*, ed. David Gooding, Trevor Pinch, and Simon Schaffer (Cambridge, Cambridgeshire; New York: Cambridge University Press, 1989), 301.

[41]Peter Louis Galison, *How Experiments End* (Chicago: University of Chicago Press, 1987), 19.

[42]Steinle, "Experiments in History and Philosophy of Science," 419.

One way in which the "scientist novice-at-the-frontier" is better off than a complete novice is through the accumulation of knowledge as systems of regularities from previous application of trial and error. This makes them more efficient users of trial and error than "complete lay novices", that is, people with no experience in the field. Just as we discovered success clues through trial and error in our search for a key, expert researchers have learned the success clues relevant to the application of trial and error to their area of research, enabling them to discard more rapidly potential trials that have no prospect of success. An analogy in our key search example would be a professional locksmith compared with the average person. The locksmith would still need to use trial and error to find the right key but training and experience at previous trials would reduce the number of keys tried in a specific lock.

Nickles's scientist novices-at-the-frontier are novices because of what they do not yet know. That is, they are novices because of their relative ignorance. Hon uses ignorance to argue for a distinction between mistakes and errors.[43] Drawing on St Augustine's observation that one cannot go wrong except through ignorance Hon distinguishes between mistakes and errors on the basis that mistakes are the result of avoidable ignorance while errors stem from unavoidable ignorance. Hon characterises so-called "arithmetic errors" like $12 \times 12 = 136$ as mistakes, the result of avoidable ignorance. He argues that in contrast Hertz's erroneous declaration that "the electrostatic and electromagnetic properties of the cathode rays are either nil or very feeble" was an error because it was the result of unavoidable ignorance caused by unrecognised problems in his experimental apparatus.[44] In making this distinction, Hon further notes that while mistakes can be identified and corrected by established checking procedures, no such procedures are available for errors.

In Hon's terms, Edison's misdirected search for a microphone resistance material was an error caused by unavoidable ignorance: his mistaken belief that the resistance of carbon varied with pressure. Applying Hon's distinction to trial and error, the error component is appropriately named, because the failures that occur (errors) are the result of unavoidable ignorance, the result of not knowing yet how to predict success. Indeed, the reason for resorting to trial and error in many instances is unavoidable ignorance. If we knew that the candidate had no likelihood of success (say from informed trial and error) we would not try it. This suggests a further reason for the frequent references to trial and error in experimental research. Trial and error is a reasonable strategy in the face of unavoidable ignorance since overcoming ignorance i.e. building knowledge, is an object of such research.

Writing of difficulties in experimental science, Hacking observes, "Many of the bugs are never understood [by any theory]. They are eliminated by trial and error".[45] Like Edison's bugs, Hacking's scientists' bugs are something unanticipated and

---

[43] Giora Hon, "Going Wrong: To Make a Mistake, to Fall into an Error.," *The Review of Metaphysics* 49, no. 1 (1995): 2.

[44] quoted in ibid., 1.

[45] Ian Hacking, "Experimentation and Scientific Realism," in *The Philosophy of Science*, ed. J D Trout, Richard Boyd, and Philip Gasper (Cambridge, Massachusetts: MIT Press, 1991), 254.

unwanted, something that does not work as it should. Bugs are failures. Hacking's scientists, like Edison, solve their bugs using trial and error, not because they are incompetent but because it is a pragmatic way to approach them.

Gooding and Nickles's observation that experimenters working at the frontiers of theory are relative novices provides a reason why even as competent researchers might resort to trial and error: they have not yet acquired relevant theory. Similarly, Edison resorted to trial and error in his search for microphone resistance material because no adequate theory for the relevant electromechanical properties of materials existed at the time. Given this, why should Tesla have believed that "a little theory and calculation would have saved [Edison] 90 per cent of the labour"? The answer lies in the kind of solutions trial and error produces. Once a trial and error solution has been found, it functions in the same way as a solution found by other means but with a significant limitation. Trial and error is a heuristic and as such is an informal method not a formal proof. Like other heuristics, trial and error may produce a solution but it provides no guarantee that it is the optimum solution or the only solution. In my key search, the first key found that opens the lock may not be the only one in the box that will open it. In more complex cases, like Edison's search for a microphone resistance material, trial and error may have yielded a solution as it appeared to do with fluff but the solution was neither the only one possible nor the best. Fluff was just a solution that fitted the success clues and success criteria identified at the time.

One of the strengths of informed trial and error is that failed trials can be used to identify new success clues and success criteria. The success clues identified through trial and error increase the user's effectiveness at identifying likely candidates while new success criteria provide the basis for a more successful solution by expanding the criteria satisfied. In some situations, there may be no need or benefit in seeking a theoretical explanation for a solution, once one solution has been found. In others, where the objective is to find a non-heuristic theoretical basis, obtaining at least one solution from trial and error, as Midgley described, can provide a starting point from which more can be discovered and from which a theory may be developed.

While trial and error may produce one solution, in situations where success criteria are quantitative or qualitative, other strategies may produce better solutions. Edison's trial and error solution to building his first electrical generators met his success criterion at the time, which was to convert mechanical energy into electricity. Later, more efficient generators were developed using theories developed in part from Edison's first generators and in part by applying mathematical and analytical techniques. Trial and error can be used to identify systems of regularities in the absence of relevant theories but when relevant theories become available, analytical and mathematical approaches may yield solutions more quickly than trial and error. They may also produce multiple solutions from which a best solution can be selected.

Since Edison's objective as an inventor was to get his invention to the patent office before competitors, any solution by any means would do, yet another reason why trial and error suited him. Once he had a solution and controlled the invention through a patent, his priorities reversed and his objective became getting the best solution rather than the first solution. For this, he could employ scientists, engineers and mathematicians who had the education and skills he lacked.

10.11   Problematic Aspects of Trial and Error

While trial and error may be the only viable way of solving some problems, it has its limitations. The first of these is that although informed trial and error may reduce the number of trials required, it does not eliminate the need to test the remaining candidates using simple trial and error. A second limitation, discussed above, is that because trial and error is a heuristic it may yield a solution but that solution may not be the only one possible or the best.

Thirdly, trial and error solutions may not provide any theoretical insights although, as illustrated by Midgley's anecdote, having found at least one solution provides a basis for seeking better solutions by other approaches. Since the knowledge derived from trial and error is directed towards finding a solution and not necessarily towards explaining why the solution works, this knowledge may be of limited value for finding a theoretical explanation.

A fourth problem is the sensitivity of trial and error to theory, particularly in relation to initial assumptions. Collingridge argues that in order to use trial and error efficiently the search should not be overly dependent on a theory because if the theory is erroneous, misapplied or misunderstood, the effort of trial and error effort is likely to be wasted.[46] Collingridge illustrates his argument by reference to Lister's introduction of carbolic acid as an antiseptic for surgery. Collingridge notes that while Lister was prompted by Pasteur's germ theory, the success of Lister's anti-septic techniques was dependent on the validity of Pasteur's theory. Edison's use of an erroneous theory about the electrical behaviour of carbon response to vibration exemplifies the consequences of the theory sensitivity of trial and error, as the erroneous belief that resistance of carbon varied with pressure misdirected Edison's trial and error experiments for months. In contrast, Edison's use of Ohm's and Joule's laws to determine a success criterion for his incandescent lamp involved no risk since the validity of these laws was well established at the time.[47] This suggests that Collingridge's claim should be qualified to a claim that trial and error is more likely to be successful if it is not dependent on theories that are new, speculative or not sufficiently developed.

For Edison, trial and error carried additional risks, the first of which was a consequence of his mastery of trial and error itself. As with any competence,

---

[46]David Collingridge, "Incremental Decision Making in Technological Innovation: What Role for Science?," *Science, Technology and Human Values* 14, no. 2 (1989).

[47]Israel, *Edison: A Life of Invention*, 170–71.

possessing it at a high level may lead a practitioner to choose it over others that might be more appropriate in the situation. In Tesla's opinion, Edison resorted to trial and error when theoretical or mathematical approaches might have proved more effective. In his account of the search for an antiknock agent, Midgley likewise highlighted the need to recognise when to abandon trial and error.

Trial and error could become a problem for Edison because of the breadth of his research interests, which led him into new fields several times in his career. While this breadth ultimately resulted in many more inventions than if he had stayed within the electromechanical field, it also carried risks. We saw earlier that experts in a field develop detailed knowledge of success clues in their field making informed trial and error more effective compared to novices. When Edison ventured into a new field, he ceased to be "an expert at being a frontier novice" and became just a novice with limited knowledge and experience relevant to the new field. In the 1880s, Edison began developing a process for magnetically extracting iron from low-grade ore. He started the venture with a high level of expertise in the application of trial and error to electromechanical technologies but none in relation to mining and heavy mineral processing. Moving into a new field as a novice neutralised the advantage of knowledge acquired previously in other fields. It increased the risk of failure as Edison demonstrated, losing many millions of dollars in his ore extraction venture.[48]

10.12   Not the First or the Last Resort

Thomas Edison fostered a popular image of himself as the simple man from Ohio who, despite limited education, was able to create remarkable inventions that those better educated could not. In some respects, his pronouncements on trial and error form part of this image building, showing Edison as a simple man using simple methods, cheerfully accepting failure but persisting until he met with success. The reality is that Edison far outstripped other equally persistent scientists and inventors because neither he nor his methods were simple.

Edison may have resorted to trial and error in desperation at times but this was not generally the case. His use of trial and error, while not always successful, was studied, well directed and relatively efficient because of his skill at exploiting knowledge from past trial and error searches. There may be an element of desperation in the use of trial and error by experimental scientists to solve experimental problems and eliminate bugs but there is no desperation in their use of trial and error in exploratory experiments. Trial and error is fundamental to this and effective experimentalists are effective in part because of their ability to use knowledge from previous applications of trial and error (success clues) to narrow the range of parameters to be tried.

---

[48]Ibid., 338–62.

Trial and error can be an effective strategy in novel situations if we face unavoidable ignorance since, as Gooding notes, in novel situations everyone is a novice. For novices, no matter how well trained or experienced, trial and error may be the only technique available for solving a problem. Used as informed trial and error, it can yield a solution more rapidly than alternatives and, regardless of how used, may be the only way to arrive at a solution. Trial and error may also, as it did for Midgley, point to a more rigorous approach or, as it did for Hacking's experimentalists, allow solutions without the need to develop a theoretical explanation. It may also provide, as it did for Edison, systems of regularities that to use in the absence of relevant theory to make predictions and find solutions to new problems.

Part IV
Reversing Edison

Chapter 11
Reverse Engineering

11.1 Reverse Engineering Is Ubiquitous

Reverse engineering is such common engineering technique that most of us have contact it daily, even if we are not aware of it. When we use an aftermarket product, it is likely to be a reverse engineered version of the manufacturer's original. When we click to accept the conditions governing their use of a piece of software, we agree not to reverse engineer it. The process of reverse engineering starts with an artefact for which the relevant functions are known.[1] Reverse engineering then creates a different artefact - not a copy - that has the relevant functions of the original. An engineer designing an aftermarket mobile phone charger would identify the physical and electrical properties of the charger made by the original equipment manufacturer and create a new charger using different components that will plug into the phone and charge it in the same way as the original. In earlier chapters we saw how when Thomas Edison set to work on a new invention, he often started with functions it should achieve and worked to create an invention that fulfilled those functions. Reverse engineering reverses this process, starting with a finished artefact and creating another that fulfils the same functions. This chapter will apply concepts developed in earlier chapters to reverse engineering and to the use of the term reverse engineering in the biological sciences.

---

[1] As in earlier chapters, the term artefact is used to refer to a human creation that is intended to serve a purpose, the purpose being expressed in terms of the artefact's functions. In considering the artefact we must consider both the object itself and the functions it serves, the object being the means to achieve them.

© Springer Nature Switzerland AG 2019

I. Wills, *Thomas Edison: Success and Innovation through Failure*, Studies in History and Philosophy of Science 52, https://doi.org/10.1007/978-3-030-29940-8_11

11.2   Reverse Engineering the Engineered

It will be useful to begin by revisiting some of the concepts developed in earlier chapters. We saw that awareness of failure, both experienced and anticipated, is crucial to creating successful artefacts. Failure gives the artefact creation process a direction and objective to the extent that the success of an artefact is a direct reflection of the approach of its creators to failure. We saw in Chap. 2 that Edison used failure to create successful inventions in a number of sophisticated ways, one being analysis of past failures identify success criteria that a successful invention should satisfy. Success criteria are directly related to the functions we want the artefact to perform or the properties we expect it to have in order to achieve those functions. For each function we need one or more success criteria to judge whether it has succeeded in achieving the function. Taken together, all success criteria for an artefact are referred to here as the artefact's success framework.

Engineering is basically a disciple of answering "how" questions. Loosely described, it is a process that starts with required functions and works out how to achieve them. This is true whether the function is something that has been done many times before, like building a bridge, or whether it is creating something that has not previously existed, commonly referred to as an invention. Like Thomas Edison developing a new invention, an engineer designing a new bridge starts with a necessarily incomplete notion of the bridge's functions and success criteria. As the process of bridge design (artefact creation) proceeds, the engineer refines and expands its functions and success criteria. The engineer might start with a broad objective: a bridge to carry vehicles across a river. This expands this by adding details of what kind of vehicles and how many, what kind of natural forces it must resist and so on. Similarly, Edison started on his path to inventing the Phonograph with the basic functions of recording and reproducing speech. As he progressed, he refined these functions, adding details about the acceptable quality of reproduction, recording time, durability of the recording and so on. As discussed in Chap. 3, this is the processes of continuous function redefinition and refinement that runs in parallel with the process of invention or engineering design. The result is that the artefact, its functions and success criteria change in themselves and in response to developments in the other functions and success criteria, often the result of actual or anticipated failures. Whether the artefact succeeds at meeting these functions is judged against success criteria so the process function redefinition and refinement implies success criteria redefinition and refinement. That is, artefact creation and success framework creation are parallel processes.

In the case of inventions, the success framework is also a novel artefact in itself but in the majority of engineering cases, like designing a bridge, many of the success criteria in the success framework are known in advance so comparatively few need be developed for the specific case. While engineers and inventors create a success framework for their artefact, the potential for identifying success criteria does not end with the artefact's creator, since other people such as users may also apply success criteria to the artefact that differ (and perhaps oppose) the designer's.

Although the designer's success framework may be the predominant one, an artefact's success framework is not fixed but is situational. Success criteria may alter over time, while different people and groups may apply different success criteria to the same artefact. Functions, and hence success criteria, are not fixed to the artefact.

In addition to success criteria used to judge the artefact's success or failure, its creator is likely to make use of success clues. These are not criteria for success but characteristics that help to separate candidates that are likely to succeed from those that are not. Many engineering rules of thumb, for example, are success clues. As with success criteria, for novel artefacts it may also be necessary to develop success clues as the artefact develops. Edison started work on the Phonograph using wax coated paper as the recording medium but found the chips of wax formed as the recording was cut into the wax fouled the surface producing extraneous noise. The absence of extraneous noise was a success criterion for the Phonograph and a success clue for this was a not producing chips. Edison's solution was to replace waxed paper with tin foil and to indent it rather than cut.[2]

Edison's approach to inventing, and his remarkable success as an inventor are intimately tied to his use of success criteria, success clues and his understanding and exploitation of the relationship between an inventions functions and the means by which they are achieved. One reason why Edison was such a prolific inventor was that he exploited the principle that one set of functions can be archived by many means and that one set of means can be used to achieve diverse functions.

Between October 1878 and April 1879 Edison applied for patents on nine different designs of incandescent lamp (light bulb). His tenth patent[3] embodied the familiar light bulb shape but the preceding designs also achieved the same function of producing light from electricity by incandescence. Edison was far from the first to do this; twenty other inventors had achieved the same function of producing light by incandescence before him.[4] What set Edison's tenth patent apart and enabled the development of electric lighting as a utility was not its physical shape but that it embodied a new success criterion: high filament resistance.

Not only did Edison invent many different means (i.e. inventions) that achieved the same function, he also used the same means (invention) to achieve different functions. On 18 October 1881 Edison was issued with three patents using essentially the same apparatus.[5] Two of these were for producing high vacuums for the manufacture of incandescent lamps but the third had a very different function, that of preserving food in vacuo. This patent used the same means, Edison's high vacuum pump, but achieved a very different function, evidence that there is not a fixed relationship between means and function.

We can now recast reverse engineering as the term is used by engineers using terminology developed from inventing. To reverse engineer something, the engineer

---

[2]TAED QP001:11.

[3]Edison. Electric Lamp.

[4]*Edison's Electric Light: Biography of an Invention*, 115.

[5]Preserving Fruit; Apparatus for Producing High Vacuums; Vacuum Apparatus [1].

starts with a fully developed artefact with a fully developed success framework and seeks to find alternative means for meeting that success framework. One engineering definition of reverse engineering used by the US Department of Defense is:

> A process by which parts are examined and analysed to determine how they were manufactured, for the purpose of developing a complete technical data package. The normal, expected result of reverse engineering is the creation of a technical data package suitable for manufacture of an item by new sources.[6]

Translating to the terminology used here, the reverse engineering process proceeds from an artefact ("part") with known functions to develop an explicit success framework ("technical data package"). The success framework developed by reverse engineering contains the functions and success criteria that the new, reverse engineered, part must have in order that it match those of the original. This success framework is then used as a specification by another manufacturer to produce a reverse engineered artefact for use in place of the original. It is also used by the Department of Defense to judge the success of the reverse engineered artefact.

In some instances the physical characteristics of the original part are crucial, for example, to enable it to fit into a larger assembly. In engineering jargon, this kind of reverse engineered part is a plug-in replacement. In other cases, physical characteristics such as external dimensions, shape and internal structure are not critical to it performing the required functions and so are not success criteria and part of the success framework. Omitting these means that the alternate manufacturer can then develop an artefact with different external dimensions and shape or different internal structure provided the resulting part complies with the success framework. Because of this, a reverse engineered part produced by an alternate manufacturer is not a copy of the original, despite being functionally identical within the limits defined in the success framework.

Figure 11.1 shows an original and two reverse engineered versions of a laptop drive. On the left is the original electro-mechanical hard disk. It employs a metallic coated disk spinning at high speed and magnets to record and retrieve data. In the centre is a solid state drive with no moving parts. Despite being very different internally, it has the same form factor as the hard disc, functions in the same way and so is a plug-in replacement to the original. On the right is another solid state drive which has the same kind of solid state internal electronics as the one in the centre but a different form factor. It is not a plug-in replacement but from the point of view of the laptop user, all three drives are functionally identical because the centre and right drives have been reverse engineered to be functionally identical to the hard disk on the left.

Edison's original carbon filament incandescent lamp, introduced in 1879, was later reverse engineered though not by that name. To reverse engineer it involved identifying those aspects that were crucial to making a functionally identical lamp.

---

[6]Department of Defense, Appendix E--Dod Spare Parts Breakout Program, (Washington: US Department of Defense, 2000), www.acq.osd.mil/dpap/dars/dfars/html/r20060412/appendix_e. htm. E-103.27.

Fig. 11.1 An electro-mechanical hard drive (left), a solid state drive of the same dimensions (centre) and a solid state drive in M.2 form factor (right)

These include being electrically compatible and being a plug-in replacement, so for example it would need to have a filament resistance of say 100 ohms, operate at 100 volts, have a transparent glass bulb of certain dimensions and be attached using an Edison Screw base. These are success criteria for the lamp but its internal structure of a carbon filament in a vacuum is not. With these success criteria in hand, the engineer could produce a functionally identical lamp. In 1911 General Electric did this when it introduced tungsten filament lamps, reverse engineered to replace Edison's carbon filament lamps. General Electric's lamps used different means (tungsten filament in an inert gas) to achieve the same functions. They also added a new success criterion: greater life.

A heart-lung machine is an example of a reverse engineered artefact where physical characteristics are not success criteria. The heart-lung machine performs the functions of the patient's heart and lungs during cardiopulmonary bypass surgery. Like the heart and lungs, it pumps blood, oxygenates it and removes carbon dioxide. It is a large piece of machinery, far too large to fit into the patient's chest, and is constructed from materials such as metal and plastic, not muscle like the original. While physically and internally very different, reverse engineering makes it appear to the patient's body as functionally identical to the organs it replaces.

The success framework of the reverse engineered artefact is usually determined by the success criteria of the original artefact but in practice will not be identical. The reason for this is that the success framework that is the output of a reverse engineering process is itself an artefact and creation of that success framework involves the application of success criteria relevant to success frameworks. In contrast to the success criteria that apply to both the original and reverse engineered artefact, success clues that applied to the original may be irrelevant or misleading for the reverse engineered artefact because of the alternative solutions involved. An important consequence is that one success clue for reverse engineering is that it produces a success framework for the new artefact containing some or all of the same success criteria as the original artefact but different success clues.

In some reverse engineering physical characteristics are critical. When the US Department of Defense purchases a reverse engineered turbine blade for a jet engine from an alternative manufacturer, it will specify its dimensional properties in considerable detail because dimensions are critical to it fitting and functioning successfully within existing engines. Within these limits, the alternative manufacturer is free to make reverse engineered turbine blades using different processes, for example casting instead of forging, or different materials, for example ceramic instead of metal.

In order to produce a reverse engineered artefact, the creators of its success framework artefact (as opposed to the reverse engineered artefact) must make decisions about which success criteria are important for the reverse engineered component artefact's success. Reverse engineers working on turbine blades would decide to include success criteria about physical characteristics whereas those reverse engineering a solid state drive may not.

Just as Edison had to identify their success criteria as he developed his inventions, the engineer reverse engineering a part must identify the success criteria relevant to it and, like an inventor, is unlikely to know all of them initially. There are many reasons for this but a particularly problematic one is errors and omissions in functions claimed by the creators of the original artefact. For example, the original part manufacturer may provide a drawing of a part being reverse engineered showing a particular dimension as 5.50 ± 0.01 mm. During reverse engineering it is found that all of the original parts measure 5.51 ± 0.01 mm at this point, the cause being a fault, such as machine wear, in the original manufacturing process. In order to fit correctly in a larger assembly (i.e. meet another success criterion), the reverse engineered parts must also measure 5.51 mm. In this case, the reverse engineer must replicate the original manufacturer's failure to meet their success criteria. The required functions are, in principle, known in advance because they have been published by the original manufacturer but have not been complied with. The task in reverse engineering in this case is to replicate the non-conformances of the original product, not the original manufacturer's dimensional success criteria.

In each of these engineering examples, most of the functions of the artefact are known in advance, that is, both the functions of the artefact and the means by which they are achieved are known. Contrast this with the experience of a person wandering through a junkyard who comes across objects that are clearly artefacts but with no obvious function. They are recognisable as artefacts because they are made of materials like as plastic or metal that do not occur naturally and, even if made from natural materials, will have signs of having been shaped by non-natural processes. While we can recognise these objects as artefacts, we may have no idea what their intended function might be. In the engineering context, there would be no sense in attempting to reverse engineer something for which there are no known functions. It might be possible to copy (as opposed to reverse engineer) an artefact but, without knowing its functions, copying would be done without relevant success criteria. Even if we were to find the purpose of the original later, we could have no confidence that the copy met all the success criteria of the original and hence no confidence that it would be a successful functional replacement for it.

11.3 Reverse Engineering the Non-engineered

We now turn to the second objective of this chapter, to examine the use of the term reverse engineering in the biological sciences. Philosopher of biology Dan Dennett claims that "literally thousands of examples of successful application of the techniques of reverse engineering to biology could be cited. Some would go so far (I am one of them) as to state what biology is, is the reverse engineering of natural systems."[7]

A search of the literature supports the first part of Dennett's claim. It shows that there are indeed "literally thousands" of references to reverse engineering in the biological sciences literature so this claim is not disputed. It is the second part, the claim that biology is reverse engineering of natural systems, that is questioned here. Dennett elsewhere claims that "biology is engineering"[8] but the two claims (biology as both engineering and reverse engineering) are not inconsistent if we take them to be two separate claims; the first that natural biological processes are analogous to engineering, and the second that biological science is analogous to reverse engineering.

A reading of biological science literature reveals two distinct techniques both referred to as reverse engineering. One technique is a direct analog of reverse engineering as discussed above. This technique uses biological models to create an artefact that has functions and properties that are close to, or identical to, the original biological entity. (I use the term *entity* here to encompass both the physical in biology, for example organs, and the non-physical, such as cognitive processes.) This technique, since it is identical to engineer's reverse engineering, I will refer to as *reverse engineering*. The other technique, the one that Dennett and many others refer to as reverse engineering, has a distinctly different objective, that of attributing previously unknown functions to biological entities by examination (Dennett refers to as inspection) of their characteristics. I distinguish this technique by the term *analytical reverse engineering*.

---

[7]Daniel Clement Dennett, *Brainchildren: Essays on Designing Minds*, Representation and Mind. (Cambridge, Massachusetts: MIT Press, 1998), 256.

[8]*Darwin's Dangerous Idea: Evolution and the Meanings of Life* (New York: Simon & Schuster, 1995), 187–228.

When engineers use reverse engineering they start with an artefact, the functions of which are known, and then develop a success framework that can be used to produce an alternative artefact that is functionally identical to the original. This approach is also used with biological systems, as in the heart-lung machine but there are many other examples such as the production of naturally occurring pharmaceuticals by industrial processes, artificial body parts, artificial intelligence (AI), even synthetic sausage casings. As with the engineering examples discussed above, the functions of the original biological entity are known (for example, the heart and lungs, animal intestines for sausage casings) and reverse engineering applied to produce them by other means.

Another technique in biology, also referred to as reverse engineering, involves the analysis of biological entities with the aim of determining their function. Plotnick and Baumiller describe it as one in which "principles of physics and engineering are directly applied to the observed structure to infer its function and faculty ... this is directly comparable to the practice of reverse engineering".[9] It is this approach that I refer to analytical reverse engineering. Plotnick and Baumiller define function as "what a feature does or how it works" and faculty as "the combination of a given form and a particular function ... what the feature is capable of doing in the life of the organism". The use of the term function to refer both to what the entity does and how it does it may be consistent with common usage but unnecessarily complicates the argument. In the present discussion, function will be used as it has previously, to refer only to what an entity does or is intended to do and not its form. The use of "capable" in the definition of faculty similarly clouds matters since, as will be argued later, just because something is capable of serving a function does not justify the claim that has that function.

Dennett claims that, "in spite of the difference in the design processes, reverse engineering is just as applicable a methodology to systems designed by Nature, as to systems designed by engineers".[10] Dennett defines reverse engineering as "the interpretation of an already existing artefact by an analysis of the design considerations that must have governed its creation".[11] Dennett's definition says nothing about reverse engineering being directed towards creating functionally identical artefacts. Rather, this kind of reverse engineering is directed towards "interpretation of an already existing artefact", implying that the artefact's functions are not known. Dennett's definition describes analytical reverse engineering.

---

[9]"Invention by Evolution: Functional Analysis in Paleobiology," *Paleobiology* 26, no. 4, Supplement (2000).

[10]*Brainchildren: Essays on Designing Minds*, 256.

[11]Ibid., 254.

11.4   Problems with Analytical Reverse Engineering

Since Dennett and "literally thousands" of others have chosen to use what they believe is an engineering technique in biology, it is reasonable to test the validity of analytical reverse engineering in what they assert is its original domain, engineering and artefact creation more generally. Applying the principles outlined earlier reveals serious problems when analytical reverse engineering as applied to the non-natural, that is, to artefacts created by humans.

11.4.1   Problem 1: One Means, Many Functions

The first problem in seeking to attribute functions to artefacts is that any single artefact (one set of means) may serve many functions. Even if we can identify one function, we cannot be sure it is the only function nor can we be sure it is the principle function of the artefact. This claim has also been advanced by Gould and Lewontin.[12] They observe that the primary function of the highly decorated spandrels of San Marco Basilica in Venice is structural. "Spandrels – the tapering triangular spaces formed by the intersection of two rounded arches at right angles are necessary architectural by products of mounting a dome on rounded arches." They then note that to the visitor to san Marco, it is the mosaic decoration on the spandrels that is their most striking feature, "The design is so elaborate, harmonious, and purposeful that we are tempted to view it as the starting point of any analysis, as the cause in some sense of the surrounding architecture. But this would invert the proper path of analysis. The system begins with an architectural constraint: the necessary four spandrels and their tapering triangular form. They provide a space in which the mosaicists worked; they set the quadripartite symmetry of the dome above."[13]

Gould and Lewontin's point is that identifying one function, the mosaic decoration, does not mean that we have identified all of the functions of the spandrels, or even the primary function. Their article spawned a debate in the over the validity of adaptationism but Gould and Lewontin's observation regarding functions was itself

---

[12]"The Spandrels of San Marco and the Panglossian Paradigm: A Critique of the Adaptationist Programme," *Proceedings of the Royal Society of London. Series B, Biological Sciences* 205, no. 1161 (1979).

[13]The function of the spandrels of San Marco is primarily to transfer the load of the dome to the columns as Gould and Lewontin assert. The weight of the dome and its shape exert horizontal and vertical forces at the perimeter of base of the dome which must be transferred to become a vertical load on the columns that support it. Since the geometry of the building means that the columns are located at some distance from the base of the dome, the spandrels perform this transfer function. It requires no specialist engineering expertise to analyse this and would be within the capacity of most engineering undergraduates. Put simply, take away the spandrels and the dome will collapse.

not new.[14] In 1947 for example, Canguilhem asserted that, "It is well known that functions in the organism are substitutable, organs are polyvalent ... for a majority of organs, which we have traditionally believed to serve some definite function, the truth is that we have no idea what other functions they might indeed fulfil".[15] Gould and Lewontin's purpose is to argue against the pervasiveness of adaptationism in evolutionary biology and in favour of an approach derived from examination of the organism as a whole, a debate that is beyond the scope of this book.

Although Gould and Lewontin point to examples of the failure of the adaptationist approach to adequately account for the data and claim that adaptationism relies on the merely plausible, they do not advance fundamental reasons for scepticism about attribution of functions. This chapter addresses this by arguing that there is a fundamental weakness in the belief that it is possible to attribute functions with any confidence based on the interpretation (Dennett's term) of artefacts and, by extension, biological entities.

The first question to be asked is how this purpose, the functions of the biological entity, are to be determined. To answer this, it is not enough to show that the entity was capable of serving a purported function. The examples given earlier in this chapter drawn from Edison's inventions show that it is possible to have multiple means for achieving the same function and further, that it is possible to use the same means to achieve very different functions. In the case of Edison's vacuum pump, without knowing what is attached to it, we cannot be sure if its function (apart from creating a vacuum) is to produce lamps or preserve fruit. We have the artefact but cannot be sure what its function might be.

In discussing his view that biology is reverse engineering, Dennett makes a claim about components (which he refers to as elements) of artefacts. He says "Elements with multiple functions are not unknown to human engineering, of course, but their relative rarity is signalled by the delight we are apt to feel when we encounter a new one".[16]

In practice the reverse of this is the norm because most artefacts have multiple functions. One does not need San Marco's spandrels to illustrate this. The walls of even a modest house have multiple functions including keeping out wind and rain, supporting the roof, providing privacy and perhaps, like the spandrels of San Marco,

---

[14]Some of their opponents disputed whether or not what Gould and Lewontin called spandrels should be more accurately termed "pendentives" or "squiches". At least one architectural dictionary defines a pendentive as "A concave spandrel leading from the angle of two walls to the base of a dome" John Fleming, Hugh Honour, and Nikolaus Pevsner, *The Penguin Dictionary of Architecture*, 3rd ed., Penguin Reference Books. (Harmondsworth: Penguin, 1980), 215. That is, a pendentive is a type of spandrel. Gould, in responding to his critics, describes the debate over terminology as "truly trivial". (Stephen Jay Gould, "The Exaptive Excellence of Spandrels as a Term and Prototype," *Proceedings of the National Academy of Sciences of the United States* 94, no. 20 (1997).

[15]Georges Canguilhem, "Machine and Organism," in *Zone 6: Incorporations*, ed. J Crary and Sanford Kwinter (New York: Zone Books, 1992), 57.

[16]Dennett, *Brainchildren: Essays on Designing Minds*, 257–58.

carrying decoration. Dennett's claim that artefacts with multiple functions are rare is not true even for the simplest of artefacts. A simple screw might, at first sight, appear to have only one function, to hold things together, but further consideration reveals more: a screw not only holds things together but, unlike gluing, allows them to be taken apart easily. Screws also have functions not related to holding things together such as hanging pictures on a wall. For their manufacturer, a function of screws is to generate profit.

In artefacts, multiple functions are the norm, not the exception as Dennett asserts. Dennett's claim does perhaps have a rhetorical purpose because it suggests (though Dennett does not assert this) that, if we believe that multiple functions in artefacts are a "relative rarity", when we identify one function of an artefact, it is likely to be *the* function of the artefact.

There is a further complication, not considered by Dennett (or indeed, by most commentators on functions in relation to artefacts). This is, as discussed previously, that the functions of an artefact are not fixed, but are situational. Consequently, we are not justified in believing that the functions identified by an artefact's creators are the artefact's only functions nor that they are fixed. Even for the artefact's creators, its functions probably changed or were modified as the artefact developed. When completed and released to the world, the creator's involvement usually ceases and artefact may acquire new functions not imagined by its creator. Box cutters may have been invented for opening cardboard boxes but they have also been used to hijack aircraft. Indeed, apart from guns and knives, most things banned from aircraft carry-on baggage are banned because they have multiple functions including potentially as weapons.

11.4.2   Problem 2: One Set of Functions, Many Means

It was noted earlier that although Thomas Edison is often identified as *the* inventor of the incandescent lamp (light bulb), many inventors produced incandescent lamps before Edison. That is, many inventors devised different means for achieving essentially the same function of producing light from electricity by incandescence. This characteristic of artefacts, that the same functions can be achieved by many means, is at the root of much inventiveness and was exploited by Edison. In discussing Edison's invention of the telephone microphone, Francis Jehl, one of Edison's associates, described 27 other designs Edison conceived but did not develop.[17] While these were not successful either technically or commercially, such alternatives to an invention that Edison had already patented had the function of inhibiting competitors who might seek to achieve the functions of Edison's

---

[17]Jehl, *Menlo Park Reminiscences*, 1, 143–54.

invention, the carbon microphone, by other means. Examination of patents generally reveals that a significant number are for novel ways of achieving the functions of existing artefacts.

Since analytical reverse engineering has, as its central objective, ascribing functions, the possibility of having many means for achieving the same functions decreases the confidence we can have about the functions attributed to an artefact by inspection. This difficulty arises because there can be differences, sometimes great, between the physical characteristics and properties of artefacts that have the same function as is the case with the computer drives in Fig. 11.1. All three achieve the same functions using different means with and without the same physical characteristics.

A counter argument offered to such examples is the assertion that, in human created artefacts, design converges. Discussing this, Dennett asserts:

> Ask five different design teams to design a wooden bridge to span a particular gorge and capable of bearing a particular maximum load and it is to be expected that the independently conceived designs will be very similar: the efficient ways of exploiting the strengths and weaknesses of wood are well known and limited.[18]

In practice a more common outcome is the reverse of this, with the five different design teams producing not one but four or five different bridge designs. Worse for the convergence argument is that if one team was asked to design a bridge to similar specifications on five different occasions they may well produce five different solutions. Engineers are trained to produce, and their clients expect, multiple design solutions to the same problem so that they and their client can select the solution that is best addresses competing success criteria. Unfortunately for clients who want standardised solutions, this tendency of engineers means it is harder to get the same design from different teams than different designs.

Submissions to design competitions that have tight functional specifications often include radically different solutions even from the same engineering design team. Figure 11.2 shows seven designs submitted by the winner of one such design completion. While it is possible to identify similarities between some of them, (A1, A2 and A3 are arch bridges, C1 and C2 are cantilever bridges), the design that was built, A3, is radically different from some of the others, notably C1 and C2.

Unfortunately for the proponents of analytical reverse engineering, engineering design does not tend to converge. The lack of convergence increases the risk of misinterpreting the differences between artefacts as evidence of different functions.

11.4.3 Problem 3: Unknown Failures

Figure 11.3 shows one of Edison's very early sketches for a telephone microphone. Figure 11.4 shows the microphone that went into production. Both use the same operating principle (sound vibrations on a diaphragm transmitted to a variable

---

[18]Dennett, *Brainchildren: Essays on Designing Minds*, 254.

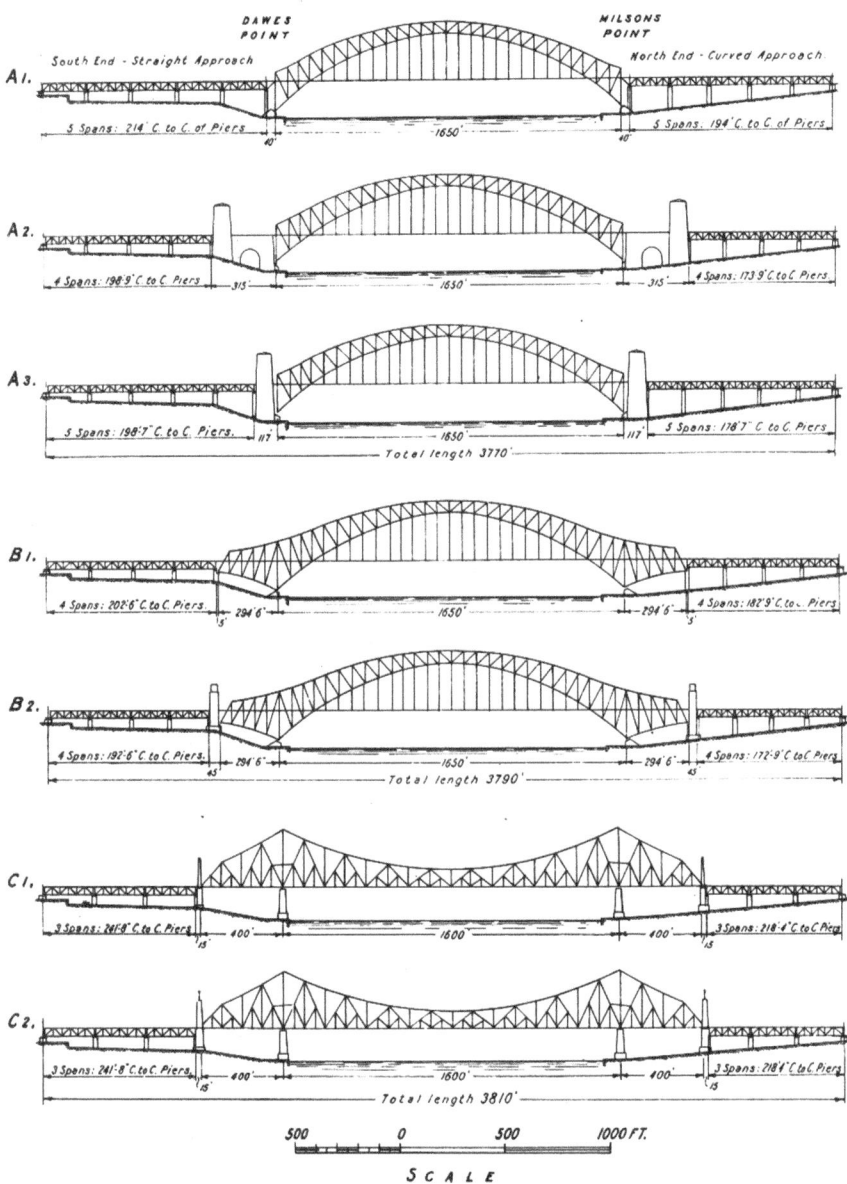

THE SEVEN DESIGNS TENDERED BY DORMAN LONG & CO., LTD.

Fig. 11.2 Alternative designs for the Sydney Harbour Bridge tendered by Dorman Long. A3 was the alternative built. (Dorman Long and Co, *Sydney Harbour Bridge*. (Middlesbrough: Dorman Long and Co., 1932))

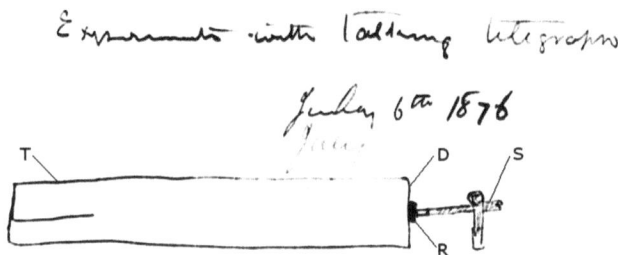

Fig. 11.3 Detail of Edison's sketch of 6 July 1876. (TAED TI2:34 notation added)

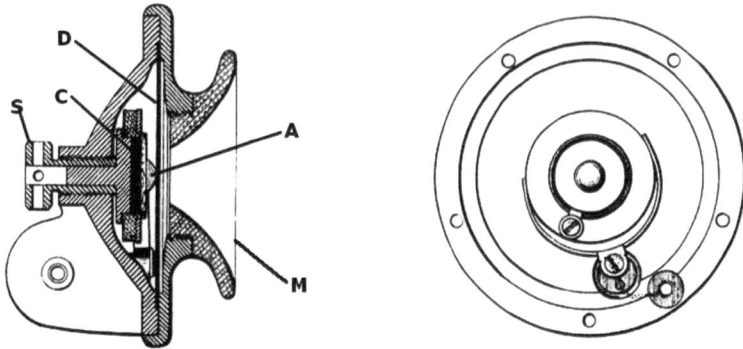

Fig. 11.4 Early production carbon microphone by Edison. (TAED TI2:490 notation added)

resistance) but there are significant differences in physical details like the shape of the speaking tube. Chapter 2 showed that having the same operating principle is not evidence that Edison progressed linearly from Fig. 11.3 to Fig. 11.4. His progress was anything but linear, and that many promising designs and components were pursued only to be subsequently discarded. Figure 11.5 shows one of Edison's intermediate designs using a rubber tube (f, Fig. 11.5) between the diaphragm (g) and variable resistance (e). The rubber tube proved troublesome so Edison replaced it with the rigid aluminium button (A in Fig. 11.4). Edison's replacement of the rubber tube with a rigid button significantly improved the performance of the invention. As there is no evidence of the tube in the final artefact there is no way for us to confidently identify the function of the metal button. Similarly, an important breakthrough in the development of the carbon microphone came when Edison constructed a device that used blocks of carbon. He then abandoned the block design but keep the carbon. Through experiments with these failed designs Edison learned how to improve his device but their contribution to its final form is made through their absence. There is nothing in the final artefact to suggest that these crucial steps were even considered, despite their significance to its development.

Fig. 11.5 An early Edison transmitter design showing a rubber tube (f) between the diaphragm (g) and resistance (D). (Edison. Speaking-Telegraphs [1])

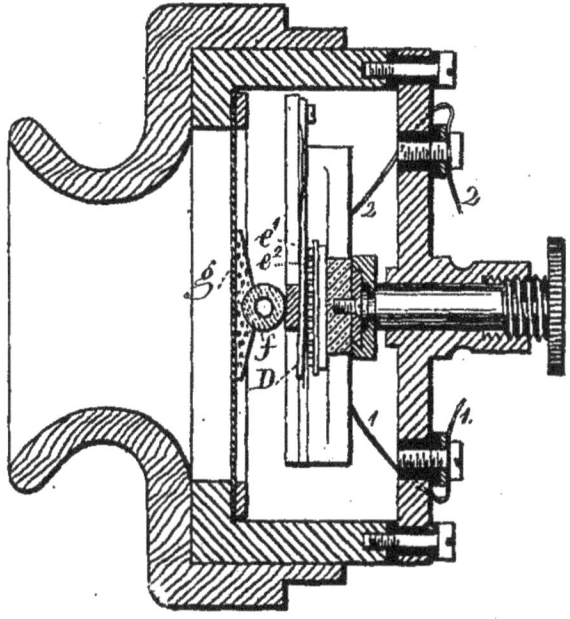

This means that the absence of something from the final form might be an oversight or was unnecessary but equally, it might also be of an intentional omission. We cannot know which by examining the final form alone, so we cannot confidently interpret absence.

11.4.4   Problem 4: Unknown Success Criteria

Problem 4 is an effect of Problem 3. When an artefact is being developed, its creator proceeds through a process of building a success framework relevant to it by identifying success criteria and success clues, learning which success criteria are, and are not, relevant, then prioritising (and often quantifying) those that are. The functions of an artefact are linked to the criteria used to evaluate the artefact's success. As discussed earlier, Edison replaced his original substituted tinfoil for waxed paper in the Phonograph to reduce noise. That is, minimising noise was a success criterion and the relevant success clue for Edison was an absence of chips on the recorded surface. Despite this criterion being crucial to the success of the Phonograph, we have no way of knowing by inspection alone that minimising noise was a success criterion or that avoidance of chips was a success clue. From the evidence of the successful artefact, tinfoil might have been Edison's first and only choice. The significance of the choice is revealed not in the artefact, but in Edison's papers.

By following Edison's development of the carbon microphone and Phonograph through his laboratory notebooks, we can see the decisions he made and the success criteria he identified as he proceeded. The notebooks also reveal that he emphasised some success criteria and not others so that the final artefact, like all artefacts, embodied a combination of functional compromises. Although Edison produced what he referred to as a "musical telephone" demonstrated at public concerts, Edison's primary objective in developing the microphone was the transmission of speech through the telephone.[19] The principal feature of the "musical telephone" was the loudness of sound it produced, not the accuracy with which it reproduced music, which was comparatively poor.

These examples emphasise that the intimate link between functions and the corresponding success criteria means that whenever we identify a function, assessing whether or not it has been achieved (whether the artefact is a success or failure in relation to the function) requires that we also have corresponding success criteria. When we examine an artefact in isolation from its developmental history, the only success criteria we can identify are those for which there is evidence in the artefact by inclusion. As with Edison's wax chip problem, without further information we cannot identify success criteria that contributed to the artefact's success if their contribution was by exclusion.

Furthermore, even with readily apparent success criteria there is often little to indicate how they were prioritised and, if they are quantitative success criteria, what the values of those quantities might have been. In some engineering cases, this can be addressed by reverse engineering the artefact so that, for example, an engineer might calculate the stresses and deflections in a completed bridge to determine its load carrying capacity. This cannot, without other evidence, reveal the quantitative load values the original bridge engineer used. A common situation is for the original engineer to start with a specific load as a success criterion but, because of uncertainties of one kind or another, apply multipliers to it (usually referred to as factors of safety) that effectively increase the original design capacity. The result is that the reverse engineered load value will, in all probability, be different from the original engineer's success criterion.

In summary, provided only with the artefact as a source of information, we cannot know all the success criteria the designer considered, their relative importance, the success criteria that resulted in exclusion of something, nor quantitative values for success criteria. This is further complicated because, as we saw earlier, functions are not inherent in an artefact, nor are they confined to the functions considered by its creator. Rather, functions are situational, varying from person to person and over time. Even with some idea of the success criteria applied by its creator to an artefact, we may know very little about the success criteria applied by others (most often users) who interact with it in some way.

---

[19]See, for example TAED MBSB1:168.

11.4.5   Problem 5: Function Without Context

Perhaps the most crucial problem and one that is both a consequence and cause of the preceding problems, is that in order to ascribe functions reliably, we should see the artefact functioning in context. That is, we cannot be sure of what it is supposed to do without seeing it do it, and in context. The following are some examples of situations where the absence of context can lead to erroneous conclusions.

The first relates to artefacts that are part of systems. Systems can be treated as a collection of components that interact and are related by a structure. Familiar examples are electrical power systems, commonly referred to as utilities. At a technological level, such a system is composed of components like generators, cables, motors and lamps that interact and are interconnected. Switching on an electric lamp causes it to draw current and this is reflected, via the interconnecting system, in increased load on the generator. Unless we have access to the whole system there is no way of understanding these interactions. Further, with only a single component, we cannot understand how its properties serve its function within the system. Edison's breakthrough in electric lighting was to recognise that his incandescent lamp required a high resistance filament. He needed this because of system interactions that were a consequence of supplying power to many buildings from a central generating station. Viewed in isolation, his lamp simply has a high resistance and there is no reason why it might not have had a low resistance as did Joseph Swann's incandescent lamps. The high resistance was crucial but created many problems for Edison. It serves no obvious function without understanding the lamp's interaction with the larger system.

Chapter 4 discussed the wider application of the notion of systems because all artefacts, parts of artefacts and assemblies of artefacts can also be treated as systems. Since most artefacts interact with something else they are components of systems. San Marco's spandrels interact structurally with the dome above and the columns below. This structural engineering interaction is one of the functions of the spandrels. Without an understating of the related systems we can understand only a fraction of the functions that an artefact might possess.

A more specialised case of artefacts as systems is that of artefacts in a set. I have an artefact that consists of two pieces of alabaster attached to each other. Viewed in isolation, it is clearly an artefact but its function is not obvious. In context, where it normally sits in front of an open door, its function is obvious. Its mass and friction with the floor prevents wind blowing the door closed. I have another identical artefact against a second door for the same reason. The function of these two alabaster artefacts appears clear: they are door stops. Before they became door stops they had another function, as bookends. The properties that made them good bookends (mass, shape, friction and aesthetics) also make them good door stops. Without seeing them in their earlier context and as a matched pair it is not obvious that they were created as bookends. They are another example of the functions of an artefact not being fixed but situational. The function of my door stops was not fixed by their maker who conceived them as bookends.

11.5   Biology as Reverse Engineering

Reverse engineering is a well-established, effective and valuable technique when applied to engineering situations and to many applications in the biological sciences where the objective is to produce something by alternative means.

When reverse engineering is used in the biological sciences to refer to the technique of analytical reverse engineering, the technique and its conclusions are questionable. Taken together, the five problems discussed above indicate that inspection of the final form of an artefact reveals only a limited amount of its functions and the process of its creation. Inspection sheds little light on the success criteria considered, the weighting or significance given to the criteria, or whether the omission of something was intentional or coincidental. Further, the artefact's form rarely reveals anything about alternatives that were discarded during development or why they were. Although crucial, they are latent in its final form. The problem that one means can have many functions, and the possibility of having many means for achieving the same function, are both potentially misleading.

Given these problems, when applied to artefacts, we cannot confidently use analytical reverse engineering to ascribe functions to Edison's inventions even though we may know quite a lot about them. If we apply analytical reverse engineering to other artefacts about which we know, less we can have even less confidence. Given that we cannot have confidence in using analytical reverse engineering to artefacts to identify functions when we know how humans create artefacts, its application to biological entities where we lack such knowledge means we cannot expect it to reliably ascribe functions to biological entities.

This is not to say that reverse engineering (as opposed to analytical reverse engineering) has no place in biology, since many have demonstrated that it has. It does not mean that we cannot determine the functions of biological entities. Rather, the conclusions of this chapter caution against reliance on analytical reverse engineering as a sole or even primary means for attributing functions to a biological entity. As with artefacts, to do so confidently we need other sources of information beyond the artefact or biological entity itself. We can tell little of Edison's processes, success criteria and intended functions by examining his inventions in isolation. Examine the inventions together with Edison's notebooks and other papers, and much can be revealed.

Chapter 12
Epilogue

12.1 Death

Thomas Edison died at home on 18 October 1931, a short distance from his New Orange, New Jersey laboratory and factory. His final illness was followed closely and his death seen by many in the United States as a national tragedy.

As in life, Edison was commemorated in death. His body lay in state in his laboratory while thousands filed past the coffin. To mark his passing, President Herbert Hoover asked the nation to switch off its electric lights for a minute on the day of his funeral.[1] In one of the more bizarre aspects of Edison's death, the air from the room in which he died was collected in a test tube and given to Henry Ford, who displayed it as Edison's "last breath" in his reconstruction of the Menlo Park laboratory next to the Ford factory.[2] Edison was also commemorated through the medium of one of his inventions, motion pictures. In 1940 Metro-Goldwyn-Mayer produced two movies based on his life. One, *Young Tom Edison* and starring Mickey Rooney, told a romanticised story of Edison's adolescence.[3] The other, *Edison the Man*, starred Spencer Tracy and presented a Hollywood version of Edison's invention of electric lighting.[4]

---

[1]Israel, *Edison: A Life of Invention*, 462.

[2]The Henry Ford. 2019. "Test Tube, "Edison's Last Breath," 1931." The Henry Ford. https://www.thehenryford.org/collections-and-research/digital-collections/artifact/225212

[3]Norman Taurog, "Young Tom Edison," (USA: Metro-Goldwyn-Mayer Studios, Culver City, California, 1940).

[4]Clarence Brown, "Edison, the Man," (USA: Metro-Goldwyn-Mayer Studios, Culver City, California, 1940).

© Springer Nature Switzerland AG 2019   243
I. Wills, *Thomas Edison: Success and Innovation through Failure*, Studies in History and Philosophy of Science 52, https://doi.org/10.1007/978-3-030-29940-8_12

12.2   Immortality

Edison's name is no longer seen on the products of the industries he founded like sound recording, electrical utilities and motion pictures but he remains with us so that, even 80 years after his death and well over a century since he was at his inventive peak, Edison continues its fascination. The *New York Times* may have been aggressively negative towards Edison in 1878, declaring that "Something ought to be done to Mr Edison, and there is a growing conviction that it had better be done with a hemp rope",[5] but the newspaper was among the first group of customers connected to his new electric lighting system in 1882. In 1922 the *New York Times* conducted a poll to find the greatest living American man awarding the honour to Edison ahead of then president, Woodrow Wilson, future president Herbert Hoover, financier John D Rockefeller, World War I commander, General John J Pershing, and industrialist, Henry Ford, among others.[6]

The fascination with Thomas Edison in the pages of the *New York Times* has not waned, as Fig. 12.1 illustrates.

Some of the peaks in Fig. 12.1 are to be expected. There is a peak in 1931 around Edison's death and also on the eve of war in 1915 when America looked to inventions to aid its defence and another in 1940 as America again faced the likelihood of war. There are also peaks on fiftieth and one hundredth anniversaries (1929 and 1979) of Edison's demonstration of his electric lighting system. This graph also reveals paradoxes in press attention to Edison. When he was at this inventive peak in the 1880s he received comparatively little attention, yet mentions of his name rise after 1890 when his patent output was falling (Chap. 9). Perhaps the most striking feature is that he continued to receive attention long after his death, to the extent that there are still years in which his name appears more often than in 1922 when he was voted the greatest living American man.

A similar pattern is evident in the Google Ngram (Fig. 12.2) which plots the number of times "Thomas Edison" appears in books in the Google Books database.

The Google Ngram shows a similar pattern to *New York Times* articles with peaks around 1931 and 1940. It also shows a surprising increase in the number of references to Thomas Edison after 1970 to the extent that in 2000 were about twice the number in 1931, the year of his death.

Edison remains significant not simply as a historic figure. In a society and era that values innovation, Edison symbolises the spirit of innovation. That he does can be attributed, at least in part, to the effort he put into promoting himself as the symbol of inventive genius, particularly after 1900. This self-promotion would not have been possible, however, without the significance of his inventions, the industries he established and their continued impact.

Despite this, Edison's career had an identifiable trajectory. In the quotation that began the Introduction, he described his inventions as initially coming with a burst

[5]New York Times, "The Aerophone."

[6]"Twelve Greatest American Men." *New York Times*, 23 July 1922, 84.

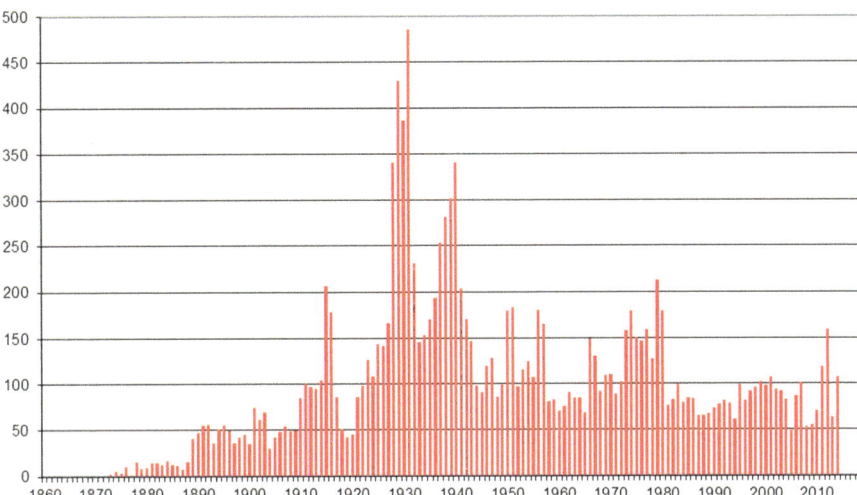

Fig. 12.1 Number of articles in the *New York Times* containing "Thomas Edison", by year. (2015. "New York Times Article Archive." New York Times. http://www.nytimes.com/ref/membercenter/ nytarchive.html)

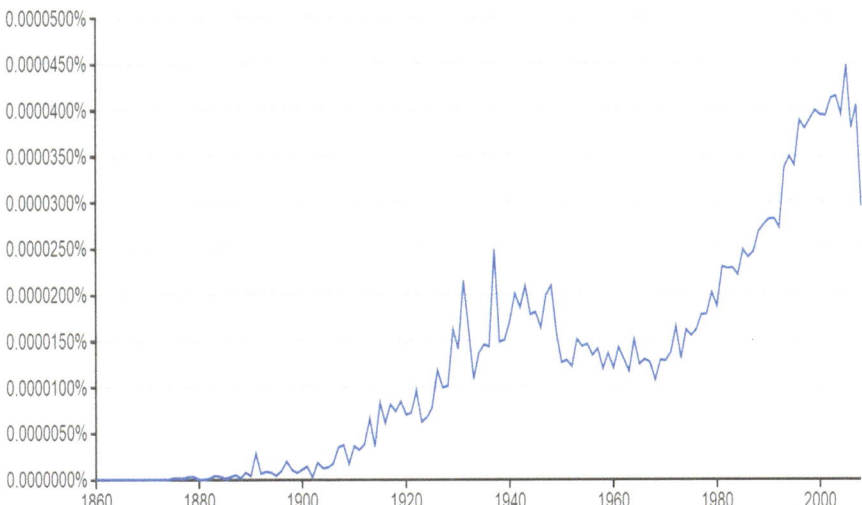

Fig. 12.2 Google Ngram of the number of times "Thomas Edison" appears in books in the Google Books database. (Google 2019. "Google Books Ngram Viewer." Google. https://books.google. com/ngrams)

followed by a long period of watching and labour. So it was with Edison's career. The inventions of his first 40 years came with a burst, peaked, and then declined over the next forty. During the period of decline Edison transformed from an impetuous and slightly eccentric inventor of miraculous things, into an industrialist and member of the American establishment, focused on preserving his commercial interests rather than seeking new radical new inventions and directions.

Edison's fame and influence was such that on at the outset of World War I in Europe, he was appointed President of the Naval Consulting Board charged with mobilising America's inventive talents to the service of national defence.[7] Edison filled the Board with what he referred to as "practical men" excluding academically trained scientists and engineers, moulding the Board's work on what he believed were his own approach to inventing. While individual members of the Board, especially Edison, created useful inventions, the Board as a whole produced little of value from the contributions of ordinary Americans. According to the historian, Thomas Hughes, the failure of Edison's Naval Consulting Board "signalled the end of the golden era of the independent inventors", the era that Edison symbolised.[8]

12.3   Legacy

At one level continued reference to Thomas Edison attests to his legacy as the archetypal American inventor. Yet his laboratory notebooks and other papers show his retrospective descriptions of his way of working made decades after his inventiveness had peaked were often inaccurate and incomplete.[9] Just as in 1891 Edison had distinguished the work of inventing from science saying, "The inventor discovers things and then the scientist steps in and tells or tries to tell what it is that has been discovered", we might paraphrase him by saying that Edison invented things and later the mature Edison tried to tell *how* he invented things.[10] He was, in his own terminology, "a practical man" who "got things done". Edison's Naval Consulting Board was modelled on Edison's beliefs about how he worked. Its failure to meet the Board's objectives (as opposed to the successes of Edison and other members as individuals), reflects Edison's mistaken beliefs about his own methods.

If Edison had left only somewhat idealised and misleading accounts of his methods, his value for understanding innovation would be limited. But he left us far more than these because his significance as in inventor meant that he also left a treasure-trove of his papers to reveal what he was doing as he was doing it.

Among those revelations is the significance of failure to his success. A little personal reflection will also reveal the extent to which we use failure to direct our daily activities. Even in common place things like getting dressed we put into effect

---

[7]Lloyd N Scott, *Naval Consulting Board of the United States* (Washington: Government Printing Office, 1920).

[8]Hughes, *American Genesis*, 15.

[9]Notable among these are Dyer and Martin, *Edison, His Life and Inventions*. and Edison, *The Diary and Sundry Observations of Thomas Alva Edison*.

[10]Chicago Daily Globe, "Arrival of Thomas A. Edison."

lessons learnt from past failures. The importance of failure increases when we seek to do something we have not been done before. The more novel the thing we set out to do, the greater the importance of seeking failure, building success criteria and overcoming them. Edison's papers reveal the ways in which we all do that.

Bibliography

Alexander, Christopher. 1964. *Notes on the Synthesis of Form*. Cambridge, MA: Harvard University Press.

American Graphophone Company Vs Edison Phonograph Works. 1896.

ASHRAE. 2009. *ASHRAE Handbook – Fundamentals*. SI ed. Atlanta: American Society of Heating, Refrigerating, and Air-Conditioning Engineers, Inc.

———. 2010. *ANSI/ASHRAE/IESNA 90.1:2010: Energy Standard for Buildings Except Low-Rise Residential Buildings*. Atlanta: American Society of Heating, Refrigerating, and Air-Conditioning Engineers, Inc.

Baile, J. 1872. *Wonders of Electricity*. Trans. John W. Armstrong. New York: Scribner Armstrong and Co.

Baldwin, Neil. 2001. *Edison: Inventing the Century*. Chicago: University of Chicago Press.

Basalla, George. 1988. *The Evolution of Technology*. Cambridge: Cambridge University Press.

Beard, George M. 1876. The Newly-Discovered Force. *The Quarterly Journal of Science, and Annals of Mining, Metallurgy, Engineering, Industrial Arts, Manufactures and Technology* April: 178–201.

Beard, George Miller. 1881. *American Nervousness Its Causes and Consequences*. New York: G. P. Putnam's Sons.

Beard, George Miller, and Alphonso David Rockwell. 1875. *A Practical Treatise on the Medical & Surgical Uses of Electricity: Including Localized and General Faradization; Localized and Central Galvanization; Electrolysis and Galvano-Cautery*. New York: William Wood & Co.

Bell, Alexander Graham. 1876. *Improvement in Telegraphy*. US patent 174,465 filed 14 February 1876, and issued 7 March 1876.

Bell, Alexander Graham, Chichester A. Bell, and Charles Sumner Tainter. 1885. *Reproducing Sounds from Phonograph Records*. US patent 341,212 filed 18 November 1885, and issued 4 May 1886.

Berger, Richard G. 1934, April. With Edison's Insomnia Squad. *Modern Mechanix* 50–52: 136–38.

Berliner, Emile. 1887. *Gramophone*. US patent 372,786 filed 4 May 1887, and issued 8 November 1887.

Blavatsky, H.P. 1877. *Isis Unveiled: A Master-Key to the Mysteries of Ancient and Modern Science and Theology*. Vol. 1. New York: J W Bouton. http://www.theosociety.org/pasadena/isis/iu-hp.htm.

Brady, Mathew. 1878. *Thomas Edison, Full-Length Portrait, Seated, Facing Front, with Phonograph*. Washington: Library of Congress. http://www.loc.gov/pictures/item/89714876/.

Brooklyn Daily Eagle. 1878, February 26. Phonograph: A Machine That Talks and Sings. *Brooklyn Daily Eagle*.

© Springer Nature Switzerland AG 2019

I. Wills, *Thomas Edison: Success and Innovation through Failure*, Studies in History and Philosophy of Science 52, https://doi.org/10.1007/978-3-030-29940-8

Brown, Clarence. 1940. *Edison, the Man*. 107 minutes. Metro-Goldwyn-Mayer Studios, Culver City, California, USA.

Canguilhem, Georges. 1992. Machine and Organism. In *Zone 6: Incorporations*, ed. J. Crary and Sanford Kwinter, 44–69. New York: Zone Books.

Carlson, W. Bernard. 1991. *Innovation as a Social Process: Elihu Thomson and the Rise of General Electric, 1870–1900*, Studies in Economic History and Policy. Cambridge/New York: Cambridge University Press.

Chicago Daily Globe. 1891, May 13. Arrival of Thomas A. Edison. *Chicago Daily Globe*.

Clark, Ronald William. 1977. *Edison: The Man Who Made the Future*. London: Macdonald and Jane's.

Collingridge, David. 1989, Spring. Incremental Decision Making in Technological Innovation: What Role for Science? *Science, Technology and Human Values* 14(2): 141–162.

Collins, Harry. 1992. *Changing Order: Replication and Induction in Scientific Practice*. Chicago: University of Chicago Press.

———. 2001, February. Tacit Knowledge, Trust and the Q of Sapphire. *Social Studies of Science* 31(1): 71–85.

Columbia Accident Investigation Board. 2003. *Columbia Accident Investigation Board Report*. Vol. 1. Washington, DC: National Aeronautics and Space Administration and the Government Printing Office. URL 7 Sept 2007: www.hss.energy.gov/deprep/archive/oversight/caib_report_volume1.pdf.

Conot, Robert E. 1979. *A Streak of Luck*. 1st ed. New York: Seaview Books.

Croffut, William. 1878, April 10. The Wizard of Menlo Park. *New York Graphic*.

Cummings, Amos Jay. 1878, February 22 A Marvellous Discovery. *New York Sun*.

Davy, Humphry. 1812. *Elements of Chemical Philosophy Part I, Vol. I*. Philadelphia: Bradford and Inskeep. http://tinyurl.com/pg33dfr.

de Dondi, Giovanni. 1974. *The Planetarium of Giovanni De Dondi Citizen of Padua: A Manuscript of 1397*. Trans. G. H. Baillie and H. Alan Lloyd. Monograph (Antiquarian Horological Society). London: Antiquarian Horological Society.

Dennett, Daniel Clement. 1995. *Darwin's Dangerous Idea: Evolution and the Meanings of Life*. New York: Simon & Schuster.

———. 1998. *Brainchildren: Essays on Designing Minds*, Representation and Mind. Cambridge, MA: MIT Press.

Department of Defense. 1998. *Mil-Hdbk-338b Military Handbook: Electronic Reliability Design Handbook*. Washington: US Department of Defense. http://www.relex.com/resources/mil/338b.pdf.

———. 2000. *Appendix E--Dod Spare Parts Breakout Program*. Washington: US Department of Defense. www.acq.osd.mil/dpap/dars/dfars/html/r20060412/appendix_e.htm.

Dorman Long and Co. 1932. *Sydney Harbour Bridge*. Middlesbrough: Dorman Long and Co.

Dyer, Frank Lewis, and Thomas Commerford Martin. 1910a. *Edison, His Life and Inventions*. Electronic Text Center, University of Virginia Library: (1998), New York: Harper and Brothers Publishers. Online text. http://etext.lib.virginia.edu/toc/modeng/public/Dye1Edi.html.

———. 1910b. *Edison, His Life and Inventions*. Electronic Text Center, University of Virginia Library: (1998), New York: Harper and Brothers Publishers. Online text. http://etext.lib.virginia.edu/toc/modeng/public/Dye2Edi.html.

Edison, Thomas A. 1872. *Paper for Chemical Telegraphs Etc*. US patent 132,455 filed 16 April, 1872, and issued 22 October, 1872.

———. 1873. Relay Magnets. US patent 141,777 filed 13 March, 1873, and issued 12 August, 1873.

———. 1875. *Improvement in Solutions for Chemical Telegraphs*. US patent 168,466 filed 26 January 1875, and issued 5 October 1875.

———. 1877a. *Phonograph or Speaking Machine*. US patent 200,521 filed 24 December 1877, and issued 19 February 1878.

———. 1877b. *Speaking-Telegraph [1]*. US patent 474,230 filed 27 April, 1877, and issued 3 May, 1892.

———. 1877c. *Speaking-Telegraph [2]*. US patent 474,231 filed 20 July, 1877, and issued 3 May, 1892.

———. 1877d. *Speaking-Telegraphs [1]*. US patent 203,013 filed 13 December, 1877, and issued 30 April, 1878.

———. 1877e. *Speaking-Telegraphs [3]*. US patent 203,015 filed 28 August, 1877, and issued 30 April, 1878.

———. 1878a. *Circuits for Acoustic or Telephonic Telegraphs*. US patent 203,019 filed 21 February, 1878, and issued 30 April, 1878.

———. 1878b. *Electric Lights*. US patent 214,636 filed 14 October 1878, and issued 22 April 1879.

———. 1878c. *Improvement in Speaking Machine*. US patent 201,760 filed 4 March 1878, and issued 26 March 1878.

———. 1878d. *Thermal Regulators for Electric Lights*. US patent 214,637 filed 18 November 1878, and issued 22 April 1879.

———. 1879a. *Electric Lamp*. US patent 223,898 filed 4 November 1879, and issued 27 January 1880.

———. 1879b. *Electro-Chemical Receiving Telephone*. US patent 132,455 filed 25 July 1879, and issued 31 August 1880.

———. 1879c. *Improvement in Dynamo-Electric Machines*. US patent 219,393 filed 10 July 1879, and issued 9 September 1879.

———. 1879d. *Phonograph*. US patent 227,679 filed 29 March 1879, and issued 18 May 1880.

———. 1880a. *Apparatus for Producing High Vacuums*. US patent 248,425 filed 29 March 1880, and issued 18 October 1881.

———. 1880b. *Magnetic Ore Separator*. US patent 228,329 filed 7 April 1880, and issued 1 June 1880.

———. 1880c. *Magnetic Separator*. US patent 248,432 filed 6 August 1880, and issued 18 October 1881.

———. 1880d. *Preserving Fruit*. US patent 248,431 filed 14 December 1880, and issued 18 October, 1881.

———. 1881a. *Vacuum Apparatus* [1]. US patent 248,433 filed 31 January 1881, and issued 18 October 1881.

———. 1881b. *Vacuum Apparatus* [2]. US patent 263,147 filed 30 August 1881, and issued 22 August 1882.

———. 1881c. *Vacuum Apparatus* [3]. US patent 266,588 filed 6 December 1881, and issued 24 October 1882.

———. 1881d. *Vacuum Pump*. US patent 251,536 filed 7 December 1881, and issued 27 December 1881.

———. 1885a. *Means for Transmitting Signals Electrically*. US patent 465,971 filed 23 May, 1885, and issued 29 December, 1891.

———. 1885b. *System of Railway Signaling*. US patent 486,634 filed 7 April, 1885, and issued 22 November, 1892.

———. 1885c. *System of Railway Signaling*. US patent 350,234 filed 7 April, 1885, and issued 5 October, 1886.

———. 1891a. *Apparatus for Exhibiting Photographs of Moving Objects*. US patent 493,426 filed 24 August 1891, and issued 14 March 1893.

———. 1891b. *Dust-Proof Journal Bearing*. US patent 472,752 filed 1 October 1891, and issued 12 April 1892.

———. 1891c. Kinetographic Camera. US patent 589,168 filed 24 August 1891, and issued 31 August 1897.

———. 1891d. *Roller for Crushing Ore or Other Material*. US patent 498,385 filed 1 October 1891, and issued 30 May 1893.

———. 1908. *Flying Machine*. US patent 491,993 filed 16 November 1908, and issued 20 September 1910.

———. 1926. *Method of Producing Sound-Record Tablets*. US patent 1,690,159 filed 5 October 1926, and issued 6 November 1928.

———. 1948. *The Diary and Sundry Observations of Thomas Alva Edison*, ed. Dagobert D. Runes. New York: Philosophical Library Inc

———. 1989a. *From Workshop to Laboratory, June 1873 – March 1876*, The Papers of Thomas a Edison, ed. Robert A. Rosenberg, Keith A. Nier, Paul B. Israel, and Melody Andrews, vol. 2. Baltimore: Johns Hopkins University Press.

———. 1989b. *The Making of an Inventor, February 1847 – June 1873*, The Papers of Thomas a Edison, ed. Reese V. Jenkins, Leonard S. Reich, Robert A. Rosenberg, Paul B. Israel, Keith A. Nier, Toby Appel, Melody Andrews, Andrew J. Butrica, and Thomas E. Jeffrey, vol. 1. Baltimore: Johns Hopkins University Press.

———. 1989c. *Menlo Park: The Early Years, April 1876 – December 1877*, The Papers of Thomas a Edison, ed. Paul B. Israel, Keith A. Nier, and Louis Carlat, vol. 3. Baltimore: Johns Hopkins University Press.

———. 1998. *The Wizard of Menlo Park, 1878*, The Papers of Thomas a Edison, ed. Paul B. Israel, Keith A. Nier, and Louis Carlat, vol. 4. Baltimore: Johns Hopkins University Press.

———. 2007. *Electrifying New York and Abroad, April 1881 – March 1883*, The Papers of Thomas a Edison, ed. Paul B. Israel, Louis Carlat, David Hochfelder, Theresa M. Collins, and Brian C. Shipley, vol. 6. Baltimore: Johns Hopkins University Press.

———. 2011. *Losses and Loyalties, April 1883-December 1884*, The Papers of Thomas a Edison, ed. Paul B. Israel, Louis Carlat, David Hochfelder, and Theresa M. Collins, vol. 7. Baltimore: Johns Hopkins University Press.

———. 2004. *Research to Development at Menlo Park, January 1879 – March 1881*, The Papers of Thomas a Edison, ed. Paul B. Israel, Louis Carlat, David Hochfelder, and Keith A. Nier, vol. 5. Baltimore: Johns Hopkins University Press.

Ellis, Keith. 1974. *Thomas Edison, Genius of Electricity*, Pioneers of Science and Discovery. London: Priory Press.

Evening Star. 1878, April 19. The Phonograph at the Capitol. *Evening Star*.

Faraday, Michael. 1933. *Faraday's Diary: Being the Various Philosophical Notes of Experimental Investigation Made by Michael Faraday During the Years 1820-1862 and Bequeathed by Him to the Royal Institution of Great Britain. Now, by Order of the Managers Printed and Published for the First Time under the Editorial Supervision of Thomas Martin M Sc*, ed. Thomas Martin, 7 vols. vol. IV, London: G. Bell and Sons Ltd.

Farrah, W.V. 1992. Reichenbach, Karl (or Carl) Ludwig. In *Dictionary of Scientific Biography*, ed. Charles Coulston Gillispie, 359–360. New York: Scribner.

Feaster, Patrick. 2007. Speech Acoustics and the Keyboard Telephone: Edison's Discovery of the Phonograph Principle. *ARSC Journal* 38 (1): 10–43.

Finn, Bernard S. 1989. Working at Menlo Park. In *Working at Inventing: Thomas a Edison and the Menlo Park Experience*, ed. William S. Pretzer. Dearborn: Henry Ford Museum & Greenfield Village.

Fleming, John, Hugh Honour, and Nikolaus Pevsner. 1980. *The Penguin Dictionary of Architecture. Penguin Reference Books*. 3rd ed. Harmondsworth: Penguin.

Forbes, B.C. 1920, October. Edison Working on How to Communicate with the Next World. *American Magazine* XC(10): 10–13.

Friedel, Robert, Paul Israel, and Bernard S. Finn. 1987. *Edison's Electric Light: Biography of an Invention*. New Brunswick: Rutgers University Press.

Galison, Peter Louis. 1987. *How Experiments End*. Chicago: University of Chicago Press.

Gates, Bill. 2007, February 25. How to Keep America Competitive. *The Washington Post*. http://www.washingtonpost.com/wp-dyn/content/article/2007/02/23/AR2007022301697.html.

Gooding, David. 1986. How Do Scientists Reach Agreement About Novel Observations? *Studies in History and Philosophy of Science Part A* 17 (2): 205–230.

———. 1990, Spring. Mapping Experiment as a Learning Process: How the First Electromagnetic Motor Was Invented. *Science, Technology and Human Values* 15(2): 165–201.

Google. 2019. *Google Books Ngram Viewer*. Google. https://books.google.com/ngrams.

Gorman, Michael E. 1994a. *Alexander Graham Bell's Path to the Telephone*. [Web site], Last Modified 07-Dec-1994. http://www2.iath.virginia.edu/albell/fgt.2.html.

———. 1994b. *Cognitive Psychology of Science*. [Web Page]. http://www.iath.virginia.edu/~meg3c/psybull.html.

Gorman, Michael E, and W Bernard Carlson. 1990, Spring. Interpreting Invention as a Cognitive Process: The Case of Alexander Graham Bell, Thomas Edison and the Telephone. *Science, Technology and Human Values* 15(2): 131–64.

Gould, Stephen Jay. 1997, September 30. The Exaptive Excellence of Spandrels as a Term and Prototype. *Proceedings of the National Academy of Sciences of the United States* 94(20): 10750–10755.

Gould, Stephen Jay, and Richard C Lewontin. 1979, September 21. The Spandrels of San Marco and the Panglossian Paradigm: A Critique of the Adaptationist Programme. *Proceedings of the Royal Society of London. Series B, Biological Sciences* 205(1161): 581–598.

Gray, Elisha. 1875a. *Improvement in Electric Telegraphs for Transmitting Musical Tones*. US patent 166,095 filed 19 January 1875, and issued 27 July 1875.

———. 1875b. *Improvement in Local-Circuit Breakers for Electro-Harmonic Telegraphs*. US patent 194,671 filed 15 February 1875, and issued 28 August 1877.

———. 1875c. *Improvement in Receivers for Electro-Harmonic Telegraph*. US patent 166,094 filed 28 June 1875, and issued 27 July 1875.

———. 1875d. *Improvement in Transmitters for Electro-Harmonic Telegraphs*. US patent 165,728 filed 28 June 1875, and issued 20 July 1875.

Hacking, Ian. 1991. Experimentation and Scientific Realism. In *The Philosophy of Science*, ed. J.D. Trout, Richard Boyd, and Philip Gasper, 247–259. Cambridge, MA: MIT Press.

Hertz, Heinrich. 1962. *Electric Waves: Being Researches on the Propagation of Electric Action with Finite Velocity through Space*. Reprint, 1893, New York: Dover. http://historical.library.cornell.edu/cgi-bin/cul.cdl/docviewer?did=cdl334&view=50&frames=0&seq=5.

Hilpinen, Risto. 1992. On Artifacts and Works of Art. *Theoria* 58: 58–82.

———. 1995. Belief Systems as Artifacts. *The Monist* 78 (2): 136–155.

Hindle, Brooke. 1981. *Emulation and Invention*. New York: New York University Press.

Hollister-Short, Graham. 2004. The Formation of Knowledge Concerning Atmospheric Pressure and Steam Power in Europe from Aleotti (1589) to Papin (1690). *History of Technology* 25: 137–150.

Hon, Giora. 1995, September. Going Wrong: To Make a Mistake, to Fall into an Error. *The Review of Metaphysics* 49(1): 1–9.

Hounshell, David A. 1975, April. Elisha Gray and the Telephone: On the Disadvantages of Being an Expert. *Technology and Culture* 16(2): 133–161.

———. 1980, February 8. Edison and the Pure Science Ideal in 19th-Century America. *Science* 207 (4431): 612–617.

Houston, Edwin J. 1871. On a New Connection for the Induction Coil. *Journal of the Franklin Institute* 61(July): 417–19.

———. 1876, January. Phenomena of Induction. *Journal of the Franklin Institute* 101(January): 59–63.

Houston, Edwin J, and Elihu Thomson. 1876, April. Electrical Phenomena. The Alleged Etheric Force. Test Experiments as to Its Identity with Induced Electricity. *Journal of the Franklin Institute* 101(April): 270–274.

Hughes, Thomas P. 1977. Edison's Method. In *Technology at the Turning Point*, ed. William B. Pickett, 5–22. San Francisco: San Francisco Press Inc.

———. 1983. *Networks of Power: Electrification in Western Society 1880–1930*. Baltimore: Johns Hopkins University Press.

———. 1986, May. The Seamless Web: Technology, Science, Etcetera, Etcetera. *Social Studies of Science* 16(2): 281–292.

———. 2004. *American Genesis: A Century of Invention and Technological Enthusiasm 1870–1970*. 2nd ed. Chicago: The University of Chicago Press.

Hughes, Thomas P. Thomas Alva Edison and the Rise of Electricity. *Technology in America: A History of Individuals and Ideas*, Carroll W. Pursell. 2nd ed, 116–128. Cambridge, MA: MIT Press, 1990.

Israel, Paul B. 1998. *Edison: A Life of Invention*. New York: John Wiley.

Jehl, Francis. 1937. *Menlo Park Reminiscences: Written in Edison's Restored Menlo Park Laboratory*. 3 vols., vol. 1, Dearborn: Edison Institute.

Jenkins, Reese V. 1984. Elements of Style: Continuities in Edison's Thinking. In *Bridge to the Future: A Centennial Celebration of the Brooklyn Bridge*, ed. Melvin Kranzberg, Brooke Hindle, and Margaret Latimer, 149–162. New York: New York Academy of Sciences.

Kline, Ronald R. 1987, April. Science and Engineering Theory in the Invention and Development of the Induction Motor, 1880–1900. *Technology and Culture* 28(2): 283–313.

———. 1989, October. Tesla and the Induction Motor. *Technology and Culture* 30(4): 1018–1023.

Koen, Billy Vaughn. 2003. *Discussion of the Method: Conducting the Engineer's Approach to Problem Solving*. New York/Oxford: Oxford University Press.

Kolodner, Janet L. 1993. *Case-Based Reasoning*. San Mateo: Morgan Kaufmann Publishers.

Kroes, Peter, and Anthonie Meijers. 2002. The Dual Nature of Technical Artifacts – Presentation of a New Research Programme. *Techné: Research in Philosophy and Technology* 6 (2): 4–8.

Kruger, Justin, and David Dunning. 1999. Unskilled and Unaware of It: How Difficulties in Recognizing One's Own Incompetence Lead to Inflated Self-Assessments. *Journal of Personality and Social Psychology* 77 (6): 1121–1134.

Lathrop, George Parsons. 1890, February. Talks with Edison. *Harpers Monthly* 80: 425–435.

Latour, Bruno. 1996. *Aramis, or, the Love of Technology*. Cambridge, MA: Harvard University Press.

Layton, Edwin T Jr. 1971, October. Mirror-Image Twins: The Communities of Science and Technology in 19th-Century America. *Technology and Culture* 12: 562–580.

Legat, V. 1862. *Reproducing Sounds on Extra Galvanic Way*. http://edison.rutgers.edu/singldoc. htm as TI2: 456–458.

McClure, J.B. 1879. *Edison and His Inventions*. Chicago: Rhodes & McClure.

Meade, Marion. 1980. *Madame Blavatsky: The Woman Behind the Myth*. New York: G P Putnam's Sons.

Midgley, Thomas Jr. 1937, February. From Periodic Table to Production. *Industrial and Engineering Chemistry* 29(2): 241–244.

Millard, Andre J. 1990. *Edison and the Business of Innovation*, Johns Hopkins Studies in the History of Technology. Baltimore: The Johns Hopkins University Press.

Mitcham, Carl. 2002. Do Artifacts Have Dual Natures? Two Points of Commentary on the Delft Project. *Techné: Research in Philosophy and Technology* 6 (2): 9–11.

Moody, Lewis F. 1944. Friction Factors for Pipe Flow. *ASME Transactions* 66 (November): 671–684.

Muirhead, James P. *The Life of James Watt with Selections from His Correspondence*, 1859. New York: D Appleton & Co. http://books.google.com/books?vid=0sKexydQgpLl_a& id=0NSFx3lV-vAC&printsec=titlepage&dq=The+life+of+James+Watt.

New York Sun. 1877, November 6. Echoes from Dead Voices. *New York Sun*. http://edison.rutgers. edu/singldoc.htm. (TAED MBSB1:77).

———. 1878, April 29. The Inventor of the Age. *New York Sun*.

New York Times. 1875, December 3. Etheric Force. *New York Times*, 4.

———. 1877, November 5. The Phonograph. *New York Times*, 4.

———. 1878, March 25. The Aerophone. *New York Times*, 4.

————. 1880, January 16. Thomas a Edison's Workshop: What a Visitor Saw and Was Told There – Some Discrepancies Noted. *New York Times*, 1.

————. 1922, July 23. Twelve Greatest American Men. *New York Times*, 84.

————. 1931, October 19. Tesla Says Edison Was an Empiricist. *New York Times*, 25. ProQuest Historical Newspapers The New York Times (1851–2003)

————. 2008, March 27. Researchers Play Tune Recorded Before Edison. *New York Times*, No. 27 March

————. 2015. New York Times Article Archive. *New York Times*. http://www.nytimes.com/ref/membercenter/nytarchive.html.

Nickles, Thomas. 1989. Justification and Experiment. In *The Uses of Experiment: Studies in the Natural Sciences*, ed. David Gooding, Trevor Pinch, and Simon Schaffer, 299–333. Cambridge/Cambridgeshire/New York: Cambridge University Press.

Oxford English Dictionary. *Phonograph, V.* [in English]. Oxford University Press. http://www.oed.com/.

Oxford English Dictionary, Third edition. 2006. *Bug, N.2.* Oxford University Press. http://www.oed.com/.

Palmieri, Paolo. 2008. *Reenacting Galileo's Experiments: Rediscovering the Techniques of Seventeenth-Century Science*. Lewiston: Edwin Mellen Press.

Perrow, Charles. 1999. *Normal Accidents: Living with High-Risk Technologies*. 2nd ed. Princeton: Princeton University Press.

Petroski, Henry. 1992. *The Evolution of Useful Things*. 1st ed. New York: Knopf.

Pettit, Michael. 2006, December. The Joy in Believing: The Cardiff Giant, Commercial Deceptions, and Styles of Observation in Gilded Age America. *Isis* 97(4): 659–677.

Pinkus, Rosa Lynn B., Larry J. Shuman, Norman P. Hummon, and Harvey Wolfe. 1997. *Engineering Ethics: Balancing Cost, Schedule, and Risk – Lessons Learned from the Space Shuttle*. Cambridge/Cambridgeshire/New York: Cambridge University Press.

Plotnick, Roy E., and Tomasz K. Baumiller. 2000. Invention by Evolution: Functional Analysis in Paleobiology. *Paleobiology* 26 (4, Supplement): 305–323.

Polanyi, Michael. 1962. *Personal Knowledge: Towards a Post-Critical Philosophy*. corrected ed, 1958. London: Routledge.

Reichenbach, Karl. 1850. *Physico-Physiological Researches on the Dynamides or Imponderables (Magnetism, Electricity, Heat, Light, Crystallisation, and Chemical Attraction) in Their Relation to the Vital Force*. Trans. William Gregory. London: Taylor, Walton, & Maberly

Rolt-Wheeler, Francis William. 1925. *Thomas Alva Edison. True Stories of Great Americans*. New York: Macmillan.

Rosanoff, Martin André. 1932. Edison in His Laboratory. *Harpers Magazine* 165 (September): 402–417.

Schiffer, Michael B. 2008. *Power Struggles: Scientific Authority and the Creation of Practical Electricity before Edison*. Cambridge, MA: The MIT Press.

Scientific American. 1877, December 22. The Talking Phonograph. *Scientific American* 384–385.

Scott, Lloyd N. 1920. *Naval Consulting Board of the United States*. Washington: Government Printing Office.

Sibum, Otto. 1995. Reworking the Mechanical Value of Heat: Instruments of Precision and Gestures of Accuracy in Early Victorian England. *Studies in History and Philosophy of Science* 26 (1): 73–106.

————. 2000. Experimental History of Science. In *Museums of Modern Science: Nobel Symposium 112*, ed. Svante Lindqvist, Marika Hedin, and Ulf Larsson, 77–86. Canton: Science History Publications/USA.

Snyder, Monroe B. 1920, March. Professor Elihu Thomson's Early Experimental Discovery of the Maxwell Electro-Magnetic Waves. *General Electric Review* XXIII(3): 208.

Steinle, Friedrich. 1997. Entering New Fields: Exploratory Uses of Experimentation. *Philosophy of Science* 64 (Proceedings): S65–S74.

———. 2002. Experiments in History and Philosophy of Science. *Perspectives on Science* 10 (4): 408–432.

Süsskind, Charles. 1964. Observations of Electromagnetic Wave Radiation before Hertz. *Isis* 55 (179): 32–42.

Taurog, Norman. 1940. *Young Tom Edison*. 86 minutes. USA: Metro-Goldwyn-Mayer Studios, Culver City, California.

Tesla, Nikola. 1887. *System of Electrical Distribution*. US patent 381,970 filed 23 December 1887, and issued 1 May 1888.

The American Whig Review. 1852. Researches of Baron Reichenbach on the "Mesmeric," Now Called the Odic Force. *The American Whig review* 15 (90): 485–501.

The Henry Ford. 2019. *Test Tube, "Edison's Last Breath," 1931*. The Henry Ford. https://www.thehenryford.org/collections-and-research/digital-collections/artifact/225212.

Thomas A. Edison Papers. 2019a. *Citing Edison Papers Documents*. [web page]. The Thomas Edison Papers, Rutgers, The State University of New Jersey. http://edison.rutgers.edu/citationinst.htm.

———. 2019b. *Edison's Patents*. [web page]. The Thomas Edison Papers, Rutgers, The State University of New Jersey. http://edison.rutgers.edu/patents.htm.

———. 2019c. *Inventions*. [web page]. The Thomas Edison Papers, Rutgers, The State University of New Jersey. http://edison.rutgers.edu/inventions.htm.

———. 2019d. *Search Method: Retrieve a Single Document or Folder/Volume*. [web page]. The Thomas Edison Papers, Rutgers, The State University of New Jersey. http://edison.rutgers.edu/singldoc.htm.

———. 2019e. *Thomas A. Edison Papers*. [web site]. The Thomas Edison Papers, Rutgers, The State University of New Jersey. http://edison.rutgers.edu/.

Thurston, Robert H. *A History of the Growth of the Steam-Engine*. New York: D. Appleton and Company, 1878. 1878 edition: http://www.history.rochester.edu/steam/thurston/1878/.

United States. Presidential Commission on the Space Shuttle Challenger Accident. 1986. *Report to the President*. 5 vols Washington, DC: The Commission. https://spaceflight.nasa.gov/outreach/SignificantIncidents/assets/rogers_commission_report.pdf.

US Congress, House of Representatives. *Resolution 269*.

US Department of Commerce. 2001. *Manual of Patent Examining Procedure*. Alexandria: US Department of Commerce. web page. http://www.uspto.gov/web/offices/pac/mpep/index.htm.

Reichenbach, Baron Karl von. 1853. *Physico-Physiological Researches in the Dynamic of Magnetism Electricity Heat Light, Crystallization, and Chemism in Their Relations to the Vital Force*. Trans. Leslie O Korth. 2nd American ed. New York: Partridge Britton.

Wachhorst, Wyn. 1981. *Thomas Alva Edison: An American Myth*. Cambridge, MA: MIT Press.

Washington Post. 1878, April 19. Genius before Science. *Washington Post*.

Weick, Karl E., Kathleen M. Sutcliffe, and David Obstfeld. 1999. Organizing for High Reliability: Processes of Collective Mindfulness. *Research in Organizational Behavior* 21: 81–123.

Weiner, Philip P. 1956, April. G M Beard and Freud on 'American Nervousness'. *Journal of the History of Ideas* 17(2): 269–274.

Willis, Martin, and Catherine Wynne, eds. 2006. *Victorian Literary Mesmerism*. C C Barfoot, Theo D'haen and Erik Kooper, Costerus, New Series., V. 160. Amsterdam: Rodopi.

Wills, Ian. 2007, July. Instrumentalising Failure: Edison's Invention of the Carbon Microphone. *Annals of Science* 64(3): 383–409.

———. 2009. Edison and Science: A Curious Result. *Studies in History and Philosophy of Science Part A* 40 (June): 157–166.

———. 2011a. *Experimental Phonograph Recording, Sample_1*. YouTube. https://youtu.be/O4wuFSDngQ8.

———. 2011b. *Experimental Phonograph Recording, Sample_2*. YouTube. https://youtu.be/WZ82DUqdCuo.

———. 2011c. *Experimental Phonograph Recording, Sample_3*. YouTube. https://youtu.be/8eGt8hk8hEw.

———. 2011d. *Experimental Phonograph Recording, Sample_4*. YouTube. https://youtu.be/e3qYQ0P0kgo.

Woodbury, D.O. 1960. *Elihu Thomson, Beloved Scientist 1853–1937*, 2nd ed.

WSWR. 1868. Art. Xxxiv- [Review of] the Medical Use of Electricity, with Special Reference to General Electrization as a Tonic in Neuralgia, Rheumatism, Dyspepsia, Chorea, Paralysis, and Other Affections Associated with General Debility. With Illustrative Cases. By Geo M Beard, M D and a D Rockwell M D. 12 Mo Pp 65. New York William Wood & Co 1867. *The American Journal of Medical Sciences* 55 (January): 236–237.

Yablonovitch, Eli. 2003, October 1. Photonic Crystals: Towards Rational Material Design. *Nature Materials* 2: 648–649.

Ziman, John, ed. 2000. *Technological Innovation as an Evolutionary Process*. Cambridge: Cambridge University Press.